Stones

Stones

Their Collection, Identification, and Uses

Second Edition

R. V. Dietrich
Central Michigan University

GEOSCIENCE PRESS
TUCSON, ARIZONA

For Allen A. Chase

ISBN 0-945005-04-0

Manufactured in the United States of America.
Printed on acid-free paper

Published by Geoscience Press, Inc.
P.O. Box 42948
Tucson, AZ 85733

9 8 7 6 5 4 3 2

Contents

Preface

This book has been prepared for students of natural history and laymen who are interested in stones. It answers such perennial questions as: Why are some stones black, others white, others red, still others different colors or combinations of colors? Why are some smooth and others rough, some nearly round and others all sorts of shapes? Why are the pretty stones I pick up at the beach so dull when I get them home? The book also explains the differences between minerals and rocks and stones. It tells about where stones may be found and how stones have been formed. It lists many ways that stones, just as they may be picked up, have been used. It includes illustrations, tables, and charts to help identify stones.

In order to make the book easier to read, most references have been omitted from the text and technical terminology has been kept to a minimum. An extensive glossary and suggested readings are given.

Most of the data and concepts presented were introduced by other geologists. To all of them who have made the results of their investigations and their deliberations available either directly or indirectly through publications, I gratefully acknowledge my indebtedness. Particular thanks is due the Danish geologist Dr. Arne Noe-Nygaard for the inspiration his *Strandsten* provided.

The original diagrams were prepared from specimens or pictures as follows:

Paula Clark: Figures 2.3, 2.4B, 2.22, 2.23, 2.25, 3.2, 3.8D, 3.9, 3.10, 3.11E, 3.12, 3.13A, 3.14, 3.15, 3.16A, 3.17B, 3.19, 3.21, 3.22, and 3.23.

Gunni Jørgensen: Figures 2.1, 2.2, 2.4A, 3.4, 3.5A, 3.6, 3.7B, 3.8A, 3.8B, 3.8C, 3.11B, 3.11D, 3.13B, 3.13C, 3.16B, 3.17A, 3.18, and 3.20.

Deborah Koehler: Figure 4.5.

Jean Parker: Figures 4.10 and 4.12

Dennis Pompilius and staff: Figures 2.16, 2.24, 2.27, 3.11C, 4.8, 4.14, 5.1, 5.2, and 5.3.

Robert Ringel: Figures 3.11A and 4.2.

Mary Schmidt: Figures 4.1, 4.3, 4.9, and 4.13.

Marie Sutherland: Figures 1.3, 2.5, 2.7, 2.8, 2.10, 2.11, 2.12, 2.14, 2.15, 2.17, 2.18, 2.19, 2.21, 2.26, 4.6, 4.7, 4.11, 4.15, and 5.5.

Michele Szok: Figures 1.2A, 1.4, 1.5, 2.3, 3.5B, 3.7A, and 4.4.

No scales are given on the sketches of the individual stones because nearly all of them and the features they represent exist on stones from pebble- to boulder-size.

Gunni Jørgensen's sketches are reproduced from *Strandsten* by kind permission of Miss Jørgensen, Dr. Noe-Nygaard, and their publisher, Gyldendal of Copenhagen, Denmark. Several of the original sketches were touched up by Robert Ringel or Michele Szok. Special thanks is given to Helge Th. Kittelsen for granting permission to reproduce his late father's (Theodor Kittelsen) famous sketch of the rubble-eating knave (page 115).

Portions of the text and a number of the tables have appeared previously in my book *Geology and Virginia*, published by the University Press of Virginia. Walker Cowan, director of that press, kindly released them for inclusion here. Four of my colleagues at Central Michigan University read and criticized all or parts of the original manuscript: David B. Jorgenson, Robert E. Kohrman, Donald E. Lahman, and Wayne E. Moore. Troy G. Cochrane, a Mount Pleasant High School student; Mary Jane Dockeray of the Blandford Nature Center of the Grand Rapids Public Museum; Joseph E. Jaworski, Science Instructor at Spring Arbor College; Gordon C. Grender, Professor of Geology at the Virginia Polytechnic Institute and

State University; and Charles B. Hunt, retired, also read and made suggestions for improving the original manuscript. In addition, Dr. Grender prepared the list of terms defined in the Glossary.

At the end of Chapter 4 in the first edition, readers were asked to inform me about additional uses of stones. The response—letters, telephone calls, conversations, notes tied to stones, and posted clippings—was gratifying. Some seventy additional uses are given in this edition. Other noteworthy changes include adding Chapter 5, Collecting Stones, and some new sketches; changing the format of the book to make it easy to carry in the field; and much updating, including the Suggested Readings.

Frances S. Dietrich, my wife, accompanied me on collecting trips, helped select the specimens for illustration, critically read the manuscript, and aided in several other ways.

Sincere thanks are given to everyone who so graciously helped me during preparation of both the original and this edition of *Stones*. . . .

April 1989 R. V. (Dick) Dietrich

List of Tables

One who values stones
is surrounded by treasures
wherever he goes.

Pär Lagerkvist, *Dvärgen*

(Den som en sten är värdeful
för mätte vara omgiven av rikedomar
var han gar.)

1

Introduction

Stones have been of interest for untold centuries. Primitive people probably picked up stones by impulse, used them, and forgot them. Eventually, however, through experience, they learned to choose the durable and readily shapable stones for weapons and tools and the beautiful or unusual stones for adornment and other decorative purposes. The beautiful ones in particular must have aroused curiosity. Why were some red, others green, purple, orange, black, or white, or even multicolored? Why did some look dull while others shone or sparkled? Why were some hard, others soft; some heavy, others light? What was the significance of all these things?

Interest led to speculation, and a number of the speculations, frequently embellished, were repeated again and again until they became legends.

One of the more curious and fairly widespread legends held that some stones, the renowned *aetites*, are capable of self-propagation. A few ancient writers even reported cases where these stones, also called *eagle stones*, were actually seen in the act of giving birth—that is, upon being broken, smaller stones emerged from their interiors.* As might be expected, these stones gained high

---

*A somewhat similar question—"Do stones grow?"—is still raised frequently by farmers and gardeners who live in regions where there is fairly deep frost penetration each winter. They raise the question because each year as they cultivate they find more and more stones,

regard and were believed by many to enhance the fertility of anyone who possessed them. In fact, the designation *eagle stone* was based on the belief that the prudent male eagle always placed at least one of these stones—generally fancied as coming from some faraway, romantic spot—in his nest in order to be sure that his mate would lay fertile eggs (see Figure 1.1). Today, we know much more about these stones. Very briefly, we know that they are clay-ironstone concretions that consist of iron-rich shells surrounding claystone cores, and that when the shells are broken, well-formed pebblelike masses of claystone do indeed emerge (see Figure 2.26).

In ancient times, some relatively common stones with more or less easily distinguishable features were widely believed to grow within certain animals; for example, Asinius in the wild ass, Kenne in the eye of a stag, and Vulturis in the head of a vulture. Furthermore, a few fairly rare stones were believed to heal wounds, to prevent sickness, or to bring good fortune. Indeed, a number of the ancient encyclopedias listed and illustrated such stones along with their postulated origins and the medicinal powers or other virtues widely attributed to them. (For an English language summary, read Chapter 4 in Adams, 1938.)

For many decades, it has been generally acknowledged that most of the legends about stones were based on fancy rather than on fact. Nonetheless, if the truth were known, it might very well be shown that birthstones and other "good-luck stones," like the "fairy stone crosses" described in Chapter 4, are modern-day holdovers or extensions of the ancient fancies.

In any case, it is clear that early humans had many reasons to accumulate different stones. In fact, anthropologists have directed attention to the clear interrela-

---

many times in recently cleared fields. In most cases, the explanation of these occurrences depends upon the fact that prior cultivation created conditions whereby the "new" stones could be frost heaved from lower levels upward into the relatively loose topsoil. The frost heaving, by the way, takes place because water expands nearly 10 percent when it freezes and the direction of easiest release is upward. However, as described in the last section of Chapter 2, some stones —that is, quasi-sandstone "stones"—have, indeed, so-to-speak grown!

Figure 1.1
An eagle carrying an aetite to its nest. (From an old woodcut in *Hortus Sanitatis*, 1497.)

tions between our emergence as human beings and our early use of stones for tools.

Today, twentieth-century people continue to search for stones. Some search for minerals, rocks, fossils, and stones for practical uses; others seek them as a hobby.

Amateur collectors, often called "rockhounds" (or "pebble pups," if youngsters), have a most interesting and enjoyable hobby. It is inexpensive because little equipment is needed. Collecting out-of-doors provides healthy exercise and relaxation. From an áesthetic point of view, countless stones possess natural beauty of form or exquisite colors unapproached even by the works of artists.

Many collectors have a feeling of adventure while collecting and a feeling of great accomplishment when they discover and acquire some unsuspected or especially fine specimen. Some gain interesting knowledge about geological processes—knowledge that increases their understanding and appreciation of the natural world. Many make new and lasting friendships with fellow collectors. Some make important contributions to the science of geology. A few make profitable discoveries.

Almost anyone can collect stones. This is true because a great variety of stones may be found within the distance of a hike or a short ride from just about anywhere, especially in glaciated regions and in areas with relatively rugged terrains. Stones occur in a great variety of places; for example, along beaches, in streams, in fields, in sand and gravel pits, in roadcuts and unpaved roadbeds, and even in gardens and yards. And the diversity of stones to be found is almost mind-boggling. It is relatively easy to find stones of nearly any color or combination of colors, stones that range from the size of a small pea to the size of a barrel, and stones of just about any imaginable shape. And it may be said rather confidently that each stone is unique.

But more of that later on. First, the meanings of a few basic terms should be understood.

A **mineral** may be defined as a *natural substance, generally inorganic, with a characteristic internal arrangement of atoms and a chemical composition and physical properties that either are fixed or vary within a definite range*. Probably one of the best ways to gain a real understanding of this definition is to examine each of its parts.

The *natural* aspect, which simply means that the substance occurs in nature, is considered a necessary condition by most mineralogists. Minerals produced in laboratories—even if identical in composition and structure to natural minerals—should be designated as *synthetic*.

The *generally inorganic* part of the definition excludes most carbon-rich materials such as coal and oil, even though for years they have constituted an extremely important "mineral resource" in terms of annual contribution to the world's economy.

The requirement of *a characteristic internal arrangement of atoms* excludes natural glass and other noncrystalline substances, and it is the most critical part of the definition. The arrangements, which are three-dimensional arrays such as the one shown in Figure 1.2, are called *crystal structures*. Crystal structures may consist of atoms of only one element or of combinations of two or more elements. For example, diamond consists of only one element (carbon); quartz, commonly called "rock crystal," consists of two elements (silicon and oxygen); and biotite, or black mica, typically consists of seven elements (potassium, magnesium, iron, aluminum, silicon, oxygen, and hydrogen). The arrangement of the atoms or ions in

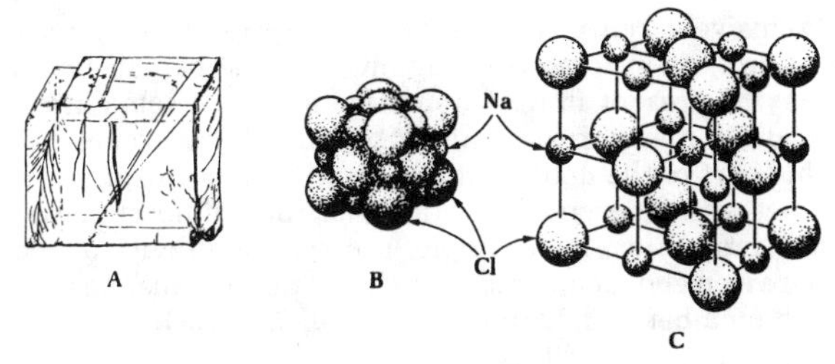

Figure 1.2
A: The mineral halite, or common table salt, typically occurs as nearly cubic cleavage fragments. *B*: Its characteristic internal structure is shown. *C*: The constituent chlorine and sodium ions have been pulled apart to show the bonds.

minerals depends upon such things as the sizes of the atoms and the way they are attached to each other. By definition, each mineral has its own unique chemical composition, crystal structure, or both. These, in turn, determine the mineral's properties: luster, inherent color, hardness, specific gravity, chemical reactivity, and so forth.

The last part of the definition requires *a chemical composition and physical properties that either are fixed or vary within a definite range*. The proportions of the constituent atoms establish a mineral's chemical composition. For example, in the common mineral quartz, atoms of silicon (chemical symbol Si) and oxygen (chemical symbol O) are present in the ratio of one silicon for each two oxygens. Consequently, quartz has a fixed composition that can be represented by the chemical formula SiO_2. And, as is true of all of the many minerals with fixed chemical compositions, quartz also has fixed physical properties—that is, all pieces of quartz, no matter where they come from, have the same specific gravity, the same optical properties, and so forth.

As also noted in the definition, however, some minerals do not have fixed compositions. Instead, their chemical compositions vary within some definite range. An example is the olivine family: $(Fe,Mg)_2SiO_4$. This chemical formula indicates that the iron (Fe) and magnesium (Mg) that combine with the silicon and oxygen (SiO_4) may be exclusively iron, as in fayalite (Fe_2SiO_4); exclusively magnesium, as in forsterite (Mg_2SiO_4); or the two elements may be present in any proportion and form some intermediate member (e.g., $Fe_{1.2}Mg_{0.8}SiO_4$). Furthermore, as this part of the definition of a mineral indicates, the olivines with different Fe:Mg ratios also have different physical properties; for example, the specific gravity of the pure iron end member is 4.39; that of the pure magnesium end member is 3.22; and that of $Fe_{1.2}Mg_{0.8}SiO_4$ would be approximately 3.89.

Each of the many hundreds of known minerals may be identified by chemical analysis, microscopic examination, x-ray diffraction, differential thermal analysis, or a combination of these methods. Fortunately, however, most of the common minerals do not have to be subjected to such sophisticated modes of identification. Instead, they may be identified on the basis of their appearances or by subjecting them to simple tests.

The general procedure is to match the properties of the unknown mineral with those listed for minerals in determinative tables. A relatively simple procedure and a determinative table for minerals that are fairly common in stones are given in Appendix A.

Rock-to-mineral relationships bring several analogies to mind: Rock is to mineral as forest is to a lonesome pine; as an exquisite tapestry is to a piece of silk thread; as a spring bouquet is to a daisy (see Figure 1.3). Certain forest-to-tree relationships clarify a number of the diverse aspects of rock-to-mineral relationships rather well: Although many natural forests contain several species of trees and most rocks contain several different minerals, some forests are made up mainly of numerous trees of a single species, just as some rocks are composed largely of many grains of a single mineral. Also, just as the trees of a forest may be of different shapes and sizes, so may the mineral grains of a rock be of many shapes and sizes. And, although some forests are similar and may be classified together, each is truly unique, just as each rock is unique.

A **rock** may be defined as *a natural solid composed of mineral grains, glass, or a combination of mineral grains and glass*. As with the definition of a mineral, this definition warrants elaboration.

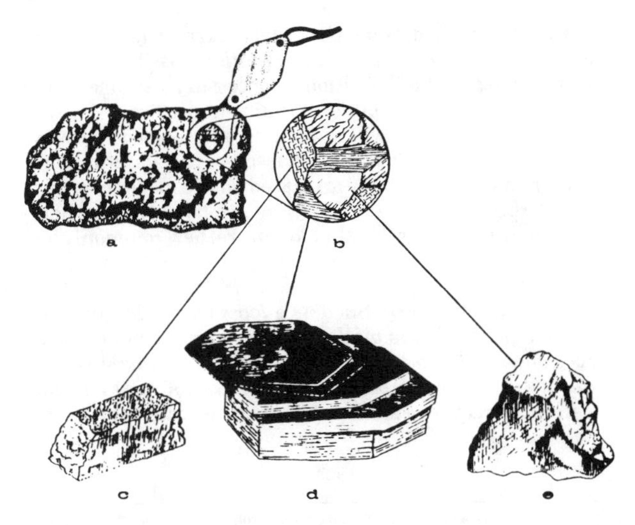

Figure 1.3
The specimen is a rock (*a*). As shown in the magnified view (*b*), this rock consists of a number of different minerals, each with its own appearance and properties. Specimens of the chief mineral constituents are feldspar (*c*), biotite mica (*d*), and quartz (*e*).

The *natural* requirement excludes materials such as concrete and slag that would otherwise fit the definition.

The *solid* aspect is highly subjective because there is a complete gradation between masses of loose fragments (such as sand) and well-consolidated, fragment-bearing rocks (such as sandstone). A simple, though certainly less than absolute, criterion is suggested by the following observation: "If you need a hammer to break it, it is a rock; if you can break it with a shovel, it is not."

The reason for using the designation *mineral grains* instead of merely "minerals" is the fact that the heterogeneous quality of some rocks is due to an arrangement involving many grains of only one mineral rather than to a composition requiring the presence of more than one mineral. *Glass* is formed when molten rock material is cooled so rapidly that the constituent atoms and ions do not have time to assume the orderly arrangements characteristic of minerals. The *combination* possibility noted in this part of the definition is necessary because some rocks consist of both mineral grains and glass.

In Chapter 3 you will see how most geologists classify and name rocks. The main classes—igneous, sedimentary, and metamorphic—are based on the modes of origin of the rocks. In Appendix C there are two tables to help you identify the rocks that occur rather commonly as stones.

A **stone** may be defined as *a loose entity (larger than a sand grain) that is made up of one or more minerals or rock materials and that has its loose aspect as a consequence of natural processes.** The essential parts of this definition require a stone to be a discrete unit; to be larger

---

**Stone* is an especially interesting word from an etymological standpoint. In the two most recently published unabridged English language dictionaries, it is noted as being a noun, a verb, or an adjective and to have twenty-nine and thirty-five different meanings (in *The Random House Dictionary of the English Language* . . . and *Webster's Third New International Dictionary*, respectively). The following poem relates a few of the rather diverse, fairly common uses.

Stonechat

A stone is a rock
  either polished or round.
A stone is a weight
  scaling four and ten pounds.
To stone a peach
  is to rid it of pit.
To stone a road
  is to make it more fit.
When Stephen was stoned,
  he died, the first martyr.
When Nero was stoned,
  he saw squirming darters.
Hailstone,
  gemstone,
    millstone,
      sand. . . .

(From *Poets of '76*, published by Quest Publications, Lavergne, Tennessee, 1977; reprinted here by permission.)

than 2 mm (~1/10 inch) in its intermediate dimension;* to be composed of one or more minerals or rock materials; and to have its free nature as a consequence of some process not directly attributable to any human activity.

So far as the *size* is concerned, different sized stones are usually given different names (Table 1.1). Although the size limits for the sub-divisions shown in Table 1.1 were set rather arbitrarily, most stones anyone might pick up would be categorized the same way by people using the scale and by those with no knowledge of it. This is fortunate because the scale is widely accepted by geologists as well as by most other scientists and engineers who deal with stones.

So far as their *composition* is concerned, some stones are made up wholly of only one or two minerals; others are made up of diverse rock materials (including natural glass); and still others consist of combinations of materials generally thought of as being either mineral or rock.

Unfortunately, the *naturally loose aspect* of many rock fragments is frequently beyond proof. Theoretically, this part of the definition stresses the felt need to distinguish between naturally loose mineral and rock fragments and just any piece of a mineral or rock. It implies that neither a piece of mineral wrested from its natural surroundings by a mineral collector nor a piece of rock broken from a bedrock exposure by a quarryman or a geologist should be called a stone; instead, each should be recognized as merely a piece or a specimen of the given mineral or rock. As a friend has so aptly put it, "Mineral specimens are often exquisite and must be carefully removed from their surroundings; rocks form cliffs, are mapped by geologists, and may be quarried or mined; stones are loose on the ground or in my garden, ready to be picked up and thrown."

To emphasize this point further, the definition of a stone subscribed to in this book not only excludes pieces of rocks and minerals removed from bedrock exposures

---

*Sand grains are for some stone collectors analogous to the micromounts of mineral collectors. The beauty of some sand grains—especially those of minerals with specific gravities so high that they sink in heavy fluids like bromoform (SG ~ 3)—is an eye-opener, especially for anyone who is looking at them for the first time through a microscope.

Table 1.1 Sizes of Stones (the Wentworth Scale)

Sizes		
mm	*inches (approx.)*	*Name*
> 256	> 10	Boulder
64–256	2–10	Cobble
2–64	1⁄10–2	Pebble*
< 2	< 1⁄10	Sand grain

*Some geologists and engineers call stones between 2 and 4 mm (~ 1⁄10 and 2⁄10 inches) *granules*.

Figure 1.4
Roundness. *A*: angular, *B*: subangular, *C*: subrounded, *D*: rounded.

by humans; it also excludes stones, as just defined, if they have been picked up and in any way reshaped by humans. In other words, neither rock materials that are commonly marketed as building stone nor artifacts are treated, as such, in this book. Furthermore, the definition excludes the discarded rock materials that make up, for example, spoil piles, tailings, and mine dumps. Therefore, it seems only prudent to note that many stone collectors extend their hobby to include rock, mineral, or fossil collecting and, when they do, they soon learn that such sites afford some of the very best collecting opportunities.

In any case, because of the difficulty in distinguishing between just any rock fragment and a true stone, some geologists have restricted the term *stone* still further to refer only to loose mineral and rock fragments that have been obviously modified by weathering or erosion since they were broken off their parent bedrock. Thus, it becomes important to describe the shapes of stones.

The two aspects of a stone's shape that are generally considered are roundness and sphericity. **Roundness** describes the sharpness of a stone's corners and edges. The most commonly used descriptive terms are *angular, subangular, subrounded*, and *rounded* (Figure 1.4). **Sphericity** is a measure of the degree to which the shape of a stone approaches the shape of a sphere. Descriptive terms such as *spheroidal, disclike, rodlike*, and *irregular* may be used (Figure 1.5).

When describing a stone, these two characteristics should always be distinguished. Roundness refers only to the presence or absence of corners and edges. Sphericity refers to the overall shape. For example, a cube is angular but has a high degree of sphericity, whereas a ball is well rounded as well as having a high degree of sphericity.

As described in the next chapter, roundness and sphericity along with a stone's surface character may be used with some success to interpret what process or group of processes have given a stone its external appearance.

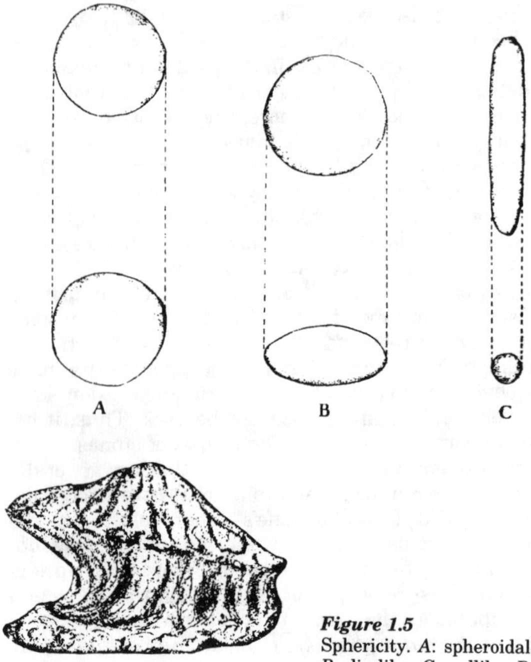

Figure 1.5
Sphericity. *A*: spheroidal, *B*: disclike, *C*: rodlike, *D*: irregular.

2

Places Where Stones May Be Found

Nearly every stone has had at least two main chapters in its history: an earlier one when its mineralogical or lithological (rock) identity was established (see Figure 3.1) and a later one during which its size, shape, surface texture, and location were acquired. The latter, the part of a stone's history that makes it differ from its kindred rocks and minerals—especially those that have not been detached from bedrock—will be considered first.

Once a bedrock unit has become exposed to the elements, fragments may be broken from it. This first step in the rock-to-stone transition almost always involves weathering. **Weathering** may be defined as the physical and chemical processes that take place as the result of exposure to the elements.

Physical weathering, also termed **disintegration**, involves the loosening and breaking up of large fragments into smaller ones without changing the chemical composition of the original substance. Some of the common causes of physical weathering are frost wedging, plant growth, and activities of animals. Frost wedging occurs when water within cracks or pore spaces of rocks or rock materials becomes so cold that it freezes. The nearly 10 percent expansion of water as it freezes is sometimes powerful enough to cause enclosing rocks to break (see Figure 2.1). Growing plants may cause similar breakage when their roots grow and expand within cracks, crevices, or other openings within rocks. Animals may disrupt rocks

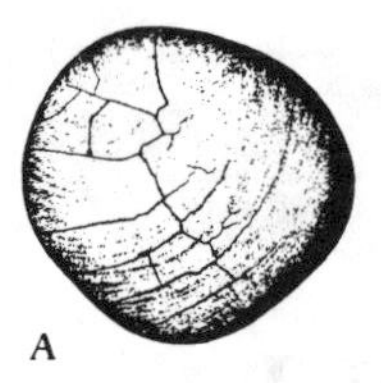

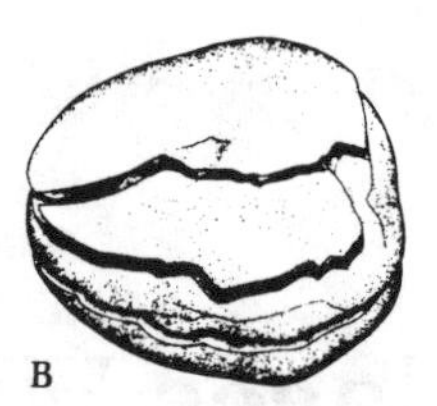

Figure 2.1
Stones may be broken when water within them freezes. *A*: A relatively homogeneous sandstone cobble. *B*: A well-layered limestone cobble.

or rock material by trampling or burrowing. In addition, in areas where exposed rocks are made up of two or more minerals that react differently to heating and cooling, disintegration may result from relatively large, fairly abrupt temperature changes.

Chemical weathering, also termed **decomposition**, involves such chemical activities as oxidation and hydration (e.g., "rusting" of iron-rich minerals). The chemical processes that take place during weathering tend to produce materials that are more stable than the original substances under the existing temperature, pressure, and moisture conditions (see Figure 2.2). Most of the chemical reactions involved in weathering require the presence of water, the universal solvent; all of the reactions appear to take place more rapidly at relatively warm temperatures than at colder ones.

Figure 2.2
Stones that have undergone chemical weathering commonly exhibit discoloration along cracks and other zones with differing water absorption tendencies.

In many places, physical and chemical weathering processes aid and abet each other. Nonetheless, in a number of regions, climatic conditions greatly favor the predominance of either physical or chemical processes. For example, cold dry climates favor physical over chemical weathering, whereas warm humid climates favor chemical weathering. Where physical weathering predominates, broken-off fragments tend to be angular; where chemical weathering predominates, the residual rubble is more likely to be rounded.

In any case, once broken from bedrock, a freed fragment becomes a stone. If the stone is not moved away from its bedrock source, it is part of the **residuum**, the unconsolidated rock and mineral weathering products that have been formed essentially where they occur. Some residual stones have their sizes and shapes modified, usually by further weathering but sometimes as the result of abrasion by windblown sand or dust. When different parts of a stone offer different resistances to the prevailing weathering or erosion processes, they often assume peculiar shapes and are said to reflect **differential** weathering or erosion (see Figure 2.3).

If a stone moves (and most stones do!), it becomes in

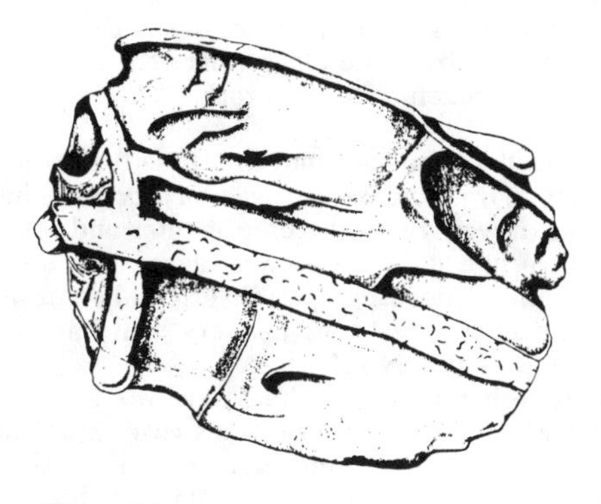

Figure 2.3
Differential weathering. This stone has ridges of relatively resistant quartz and granite that stand out in relief above the less resistant host material.

the language of the geologist colluvial, alluvial, glacial, lacustrine, or marine—the designation depending upon which natural agent of erosion is responsible for the stone's transport or deposition. **Colluvial** refers to materials moved downslope in response to gravity; **alluvial** or **fluvial**, refers to materials transported and deposited by running water; **glacial**, to those transported by glaciers; **lacustrine**, to those transported by lakes; and **marine**, to those transported by the seas and oceans. A parallel adjective for wind processes is **aeolian**, or **eolian**. In addition, stones may be transported and deposited by plants, animals (including man), and floating ice.

In many areas, especially in those once covered by glacial ice, a large majority of the stones have been transported by two or more agents. A general sequence for a well-rounded stone one might find in a gravel pit in such a region is as follows:

1. breaking of the initial fragment off the parent bedrock;
2. downslope transfer into a stream;
3. intermittent transport along a stream bed into a lake;
4. to-and-fro movement in breaking waves along the lake shore;
5. incorporation into an advancing glacier; and
6. deposition in a gravelly moraine, perhaps several tens of kilometers from the place where the original fragment was broken off the bedrock.

Historically, from the time the stone first started its downslope movement until it was incorporated into the glacial ice, it would have become smaller and smaller, it would have had its corners and edges more and more rounded, and, depending upon its original shape and its fracturing tendencies, it might very well have become more and more nearly spherical.

On the other hand, as the stone was subsequently transported by the glacier—in some cases much farther and for a much longer time than by any of the other transporting agents—the stone would probably have undergone little if any further change in character unless it was one of the few glacially transported stones that was scraped along the bedrock surface over which the glacier moved.

In any case, a stone may have both its size and its shape changed during transport. The changes depend upon processes usually referred to as abrasion, attrition, and solution. **Abrasion** is the mechanical wearing down produced as the result of impact or friction; **attrition** is the reduction of the tools of abrasion caused by impact breakage; **solution** is the chemical dissolving of a substance by a fluid. Abrasion and attrition take place whenever a collision between transported fragments or between a transported fragment and a more or less immobile target (for example, bedrock or a streambed boulder) involves enough energy to cause breakage. Abrasion and attrition may, of course, be complementary activities. Solution is rather ineffective for all but the fairly rare stones made up largely of minerals that may be readily dissolved. Each of the changes reduces the stone's size. Some of the changes modify the stone's shape only slightly; others alter it drastically.

Several things may control a stone's shape (see Table 2.1). Most important are the stone's makeup, the kinds

Table 2.1 Common Characteristics of Stones Transported by Diverse Agents

Agent	*Characteristics*	*Figure References*
Chemical weathering	Ellipsoidal; rounded; onionlike layering; stained (e.g., rust-colored) exterior.	2.2 and 2.14
Physical weathering	Blocky, angular; has unmodified fractures typical of the constituent rock type.	1.4A
Gravity (colluvial)	Any shape; may resemble either of preceding types.	
Glacial	Snubbed corners; may have one or more facets (relatively flat surfaces) that are striated.	2.4A
Water (alluvial, lacustrine, marine, and glacio-fluvial)	Diverse ellipsoidal shapes; rounded; mattelike exteriors.	1.4C & 1.4D; 3.4–3.15; and 3.18–3.23
Wind (aeolian)	Keeled, grooved, pitted, etc.; mattelike to polished exteriors.	2.4B

of weathering or erosion processes involved, and the length of time the stone was actually undergoing the rigors of change. A stone's mineralogical composition determines how it will react to both chemical weathering and most physical processes. The way the mineral grains are fitted together generally determines the way a stone will break. Slate, for example, tends to break into flat stones because it is made up of platy mineral grains arranged essentially parallel to each other.

That different agents of erosion may have different degrees of effectiveness is quite apparent when it is considered that, for instance, a stone is likely to get banged up a lot more while moving in water than in glacial ice. And, of course, even a single agent may range greatly in its effectiveness; for example, a fast-flowing stream may cause much more potent impacts than a slower-moving stream.

The other main factor controlling a stone's shape, the time element, is actually a measure of such things as the number of effective collisions. Unfortunately, it is often assumed that the time element correlates directly with the distance traveled. This is not necessarily true. For example, transport over a short distance with many collisions would cause more changes than transport over a longer distance with fewer collisions. Nevertheless, the generalization is a good rule of thumb in that the farther a stone is carried, the more times it is likely to get hit and, consequently, the more modifications its shape will tend to sustain.

In some cases, a stone's surface texture provides the best indication of the agent responsible for its erosion. Among the more commonly employed descriptive terms are *polished, dull, frosted,* and *striated*. On the basis of their surface textures, stones eroded by wind action can generally be distinguished from stones modified while being carried by a moving glacier, and both are different from stones fashioned within moving water (see Figure 2.4).

Unfortunately, there appears to be no definite way to distinguish between individual stones formed in different aqueous environments; that is, between stream-, lake-, and marine-eroded stones. If, however, a fairly large number of stones from a given deposit is considered, it

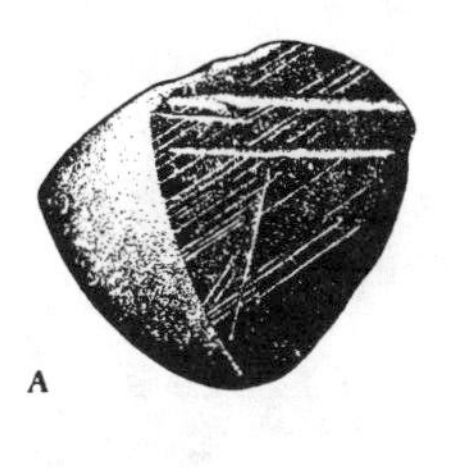

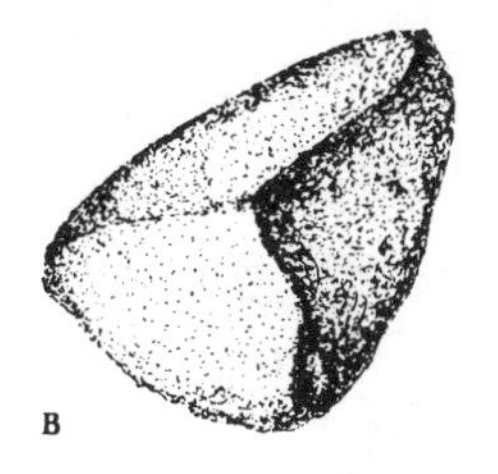

Figure 2.4
A: Glacially transported stone with a striated, nearly flat surface. *B*: Wind-eroded stone (ventifact) with polished surfaces and intervening keels.

may be possible to suggest in which of the environments the stones were shaped. Two examples may be cited: Several studies have indicated that beach stones are, as a group, flatter than stream pebbles; other studies have shown that beach deposits that consist largely of pebbles or cobbles tend to contain less sand than comparable stream deposits.

Because there are several good bases for interpreting the way stones have gained their shapes and surface characteristics, places where different kinds of stones are likely to be found can be predicted. The following are brief descriptions of different kinds of stone-containing deposits. Beach, stream, and glacial deposits are described first because they afford the vast majority of stone-collecting opportunities.

BEACH DEPOSITS

Beaches are generally defined as shore areas that extend from the low water line to the area of change in topography or vegetation. There are literally hundreds of thousands of kilometers of lake and ocean beach areas that are covered by sand and gravel. As a group, these shoreline deposits afford some of the best stone-collecting opportunities in the world (see Figure 2.5).

Many beaches are covered by sand; others, mostly by stones. Some beaches are covered by stones of a single kind of rock, commonly only slightly rounded and of many sizes; others are covered by stones made up of many kinds

Figure 2.5
Some beaches are covered with stones.

of rocks, typically well rounded and all of the same general size. Along this line, on a recent collecting trip to two islands—one in Lake Huron, the other in Lake Michigan—we found one island to have well over 99 percent of its beach stones derived from a nearby cliff and the other island to have stones representing more than fifty different igneous, sedimentary, and metamorphic rocks, most of which occur as bedrock several tens of kilometers north of the island (i.e., the stones have been transported by glacial ice into the general area where they now occur). And, as might be expected, the stones at the first island were less well size-sorted and many were less rounded than the stones on the second island. Nevertheless, almost all of the stones at both localities showed evidence of having undergone abrasion during their to-and-fro transport by the almost continuously moving lake waves.

Close analyses of waves have shown that wave forms have depth below, as well as height above, the general level of the surface. As a consequence, wherever a wave form reaches water too shallow to contain its total dimensions, the wave steepens and then curls over and breaks. In breaking, a turbulence is set up and stones and other materials that are on the bottom are lifted and moved. Collisions are common during transport; the rattling or castanetlike sounds you sometimes hear along pebble and cobble beaches are from such collisions. Additional collisions occur as the stones and sand flow back off shore (downslope) in the undertow. As a result of the impacts, the stones' shapes, sizes, and surface textures are changed. With careful observation, these processes can be seen in action along most pebble and cobble beaches.

Collecting stones along a beach is generally great fun; among other things, the stones there are wet and thus exhibit their colors just as if they were polished. For some people, however, this leads to disenchantment because "most of the pretty stones I pick up at the beach look so dull when I get them home." Fortunately the apparent change can be easily remedied by means described in Chapter 5.

Where are the stone beaches? There are literally hundreds of them; it is primarily a matter of looking. Some are rather dramatically pointed out by names on maps; for example, Stoney Point, Gravel Island, Jasper Beach, and Agate Harbor. Others—like Land's End, near San Francisco; Whitefish Point, Michigan; Geneva-on-the-lake, Ohio; various coves along the Oregon coast; Snowplow in the Bonneville Basin, Utah; North Hampton Beach, New Hampshire; and Corney Brook campground, on the Gulf of Saint Lawrence, in the Cape Breton Highlands National Park, Nova Scotia—are well known to collectors. And most collectors are more than willing to tell new converts to their hobby about good collecting localities. In addition, a walk or ride along almost any lake or ocean shore, especially in glaciated regions, will turn up so many collecting areas that you may find it hard to believe. Perhaps somewhat surprisingly, if your interest is in stones with peculiar shapes, tropical marine beaches near coral reefs are especially productive.

Along with the present-day beaches, there are many miles of "fossil beaches" (i.e., ancient beach deposits) that are currently well up on dry land. These deposits were made when the associated bodies of water were at higher levels with respect to the land and thus had different sizes and shapes than they do today. Especially good examples occur around Great Salt Lake, Utah; surrounding several of the closed basins of the Basin and Range Province; and here and there in the northern Great Lakes region.

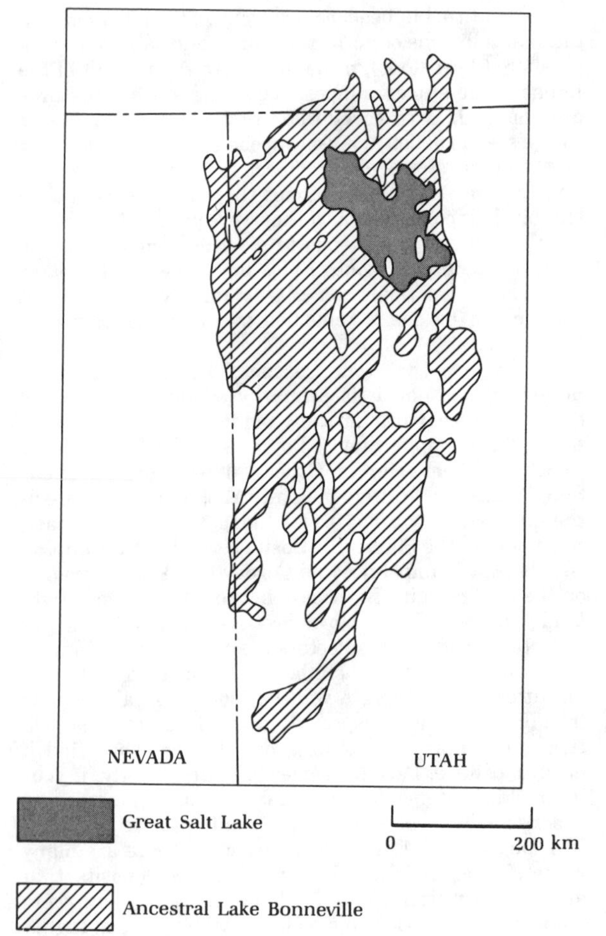

Figure 2.6
Map showing comparative areas of present-day Great Salt Lake and ancestral Lake Bonneville.

The most obvious deposits in the Great Salt Lake area are on top of prominent terraces, which look like giant steps, on the sides of the mountains that rim the basin. A few of the ancestral lakes have been named. For example, that of Great Salt Lake is Lake Bonneville, which had more than six times the area of the present-day lake (Figure 2.6).

The high-level beach deposits of the northern Great Lakes region are also associated with terraces, which are prominent wave-cut banks. These banks occur where the ancestral shore zones have undergone uplift in response to the *isostatic rebound* that has taken place since retreat of the continental ice sheet that covered the area during the Great Ice Age. In essence, this rebound is a postglacial reestablishment of the preglacial elevations that were depressed when the earth's crust was under the weight of the thick ice. Although it is a much, much slower process, the effect on the earth's crust in response to the addition and removal of large volumes of ice is analogous to the effect that results from pressing one's thumb into a rubber ball and then releasing that pressure.

Many "fossil beach deposits" contain stones that collectors would be proud to have in their collections. Locations of several of the more prominent former shorelines are indicated on appropriate state and regional maps.

STREAM DEPOSITS

Many of the world's several million streams have stones as part of their streambeds (see Figure 2.7). Several of the activities of running water have been analyzed mathematically; a few have been resolved into equations or "laws." As an example, it has been shown that a stream's *competence*, or the size of the largest fragment of a given specific gravity that the stream can move, is directly proportional to the square of the velocity of the water. For example, by merely tripling the velocity, a fragment with a nine times larger effective diameter may be transported. This accounts for a stream's ability to carry much larger fragments in its relatively steep reaches than in its flatter portions and also during its rapid-flow flood stages than during its slower "normal" flow stages. It also accounts for the formation of *placers* and "fossil placers,"

Figure 2.7
Streambeds may offer fine stone-collecting opportunities.

most of which are stream deposits that contain noteworthy concentrations of heavy minerals such as gold nuggets and some of the heavier gemstones.*

In any case, whenever a stream's competence is sufficient, stones may be carried in suspension, rolled, or tumbled along and, during that transport, have their shapes, sizes, and surface textures modified. Just as along beaches, changes are caused by the sandblast effect as passing sand and silt grains pelt motionless or slowly moving larger stones, by the grinding of stones against

---

*Note, for example, the following specific gravities:

Corundum (ruby and sapphire)	4.0–4.1
Spinel	3.55
Diamond	3.50–3.53
Topaz	3.49–3.57
- - - - - - - - - - - - - - - - - - -	
Feldspar (the common ones)	2.55–2.69
Quartz	2.65
Agate	2.57–2.64

each other or against bedrock, by breakage, and (rarely) as a result of chemical solution.

Two kinds of stones often included in collections are found almost exclusively in stream deposits: Most nearly spherical stones are formed in potholes in streambeds (see Figure 2.8) and a large percentage of the stones that bear percussion marks (crescentic cracks) appear to have been formed in rampaging streams.

A third type of stream-eroded stone is also eagerly sought by some of the more sophisticated stone collectors. What they try to find are pairs of adjacent stones that roughly fit together. Such pairs of stones are sometimes fashioned in swift-flowing streams, where some of the more or less entrapped boulders move with the currents

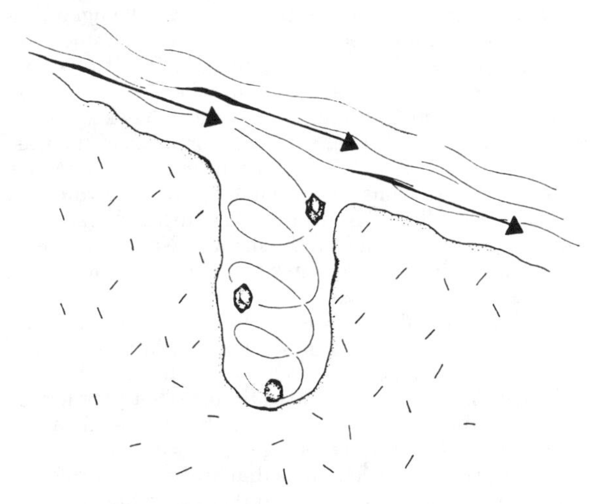

Figure 2.8
Stones whirled around in eddies may abrade holes, called potholes, in the bedrock or even in large boulders of their streambeds. If the stones that cause the abrasion are relatively homogeneous, they tend to be worn down to nearly perfect spheres. (Diagram modified and redrawn after Olof Ängeby, 1951, *Pothole Erosion in Recent Water-falls*. Lund (Sweden) Studies in Geography.)

and jostle against each other, thereby abrading each other so that their modified shapes exhibit mutual accommodation.

In many streams, the coarseness of the transported stones decreases from the headwaters toward the mouth. On the other hand, in streams flowing in unconsolidated glacial deposits, there may be several sporadic sources of large fragments along a stream's course and thus the arrangement of sediments, especially according to size, may be extremely complex. Nonetheless, some possible collecting spots are predictable: For example, wherever a stream has a falls or a sudden change from a relatively steep to a lower gradient, stones may be abundant; and wherever a stream has gravel bars, the bars may be made up chiefly of pebbles or small cobbles.

Some of the finest stones in the Central Michigan University collection were collected from a number of streambeds: where rivers come out of the Blue Ridge onto the Piedmont in central Virginia; where a large creek comes out of the Adirondacks into the Mohawk Valley in Central New York; where streams come out of the Cascades near Eugene, Oregon; and near the mouths of a couple of the streams flowing off Mona Loa, Hawaii. Still others were collected from gravel bars in braided streams on Iceland and on the south island of New Zealand. In the future, others will certainly be added from streambeds in other parts of the world.

A different but equally interesting kind of deposit that consists largely of alluvial stones is rather common in semiarid to arid regions such as the Basin and Range Province of the western United States. That province, as the name implies, is characterized by staggered ranges of mountains with intervening (or, in some places, surrounding) relatively flat areas that are generally referred to as *basins*. In response to weathering and erosion, the ranges are worn down and the basins are filled up with the resulting detritus. In some places, nearly flat to gently sloping bedrock surfaces have been formed along the bases of the receding mountains. These surfaces, generally called *pediments*, are predominantly bare bedrock over which the stones and other detritus are transported. Here and there, however, the pediments are blanketed by a veneer of stones. In other places, large quantities of stones and other detritus have been deposited directly in front of the mountains. These deposits, which typically fan out into the basins from the canyon mouths out of which the supply streams flow, are called *alluvial fans*. In several places, adjacent fans have coalesced to form essentially continous aprons of detritus. These are generally termed *bajadas*. Many pediments and nearly all alluvial fans and bajadas afford excellent opportunities for collecting stones. In fact, these deposits have served as the source for most of the stones used in the well-known desert landscaping of the southwestern United States.

Stream-eroded stones are also common in glacial deposits. Some are previously stream-fashioned stones that have not been modified since they were incorporated in and transported by glacial ice; others owe their shapes to the activities of glacial meltwater streams, generally referred to as *glacio-fluvial*. These latter stones are treated in the next section.

GLACIAL DEPOSITS

As glaciers advance, they incorporate stones and other loose rock materials that they encounter, transport and sometimes modify those materials, and finally deposit them. Glaciation during the last approximately two million years, commonly referred to as the Great Ice Age, was responsible for the presence of the great variety of stones that now occurs in many areas throughout the world (see Figure 2.9). The stones in the central part of Michigan's Lower Peninsula afford a fine example; the almost innumerable rock types present include igneous and metamorphic rocks carried in from Canada and many different sedimentary rocks that have come from bedrock areas in the eastern part of the state's Upper Peninsula and the northern part of its Lower Peninsula. Fifty to a hundred different kinds of rocks can be collected rather easily at nearly all of the gravel pits that are located throughout the area.

As a group, glacially deposited materials are called *drift*. Those that are unsorted and typically more or less unstratified are called *till*; those that are well sorted and layered are termed *stratified drift*.

Nearly all glacial till includes stones that have been

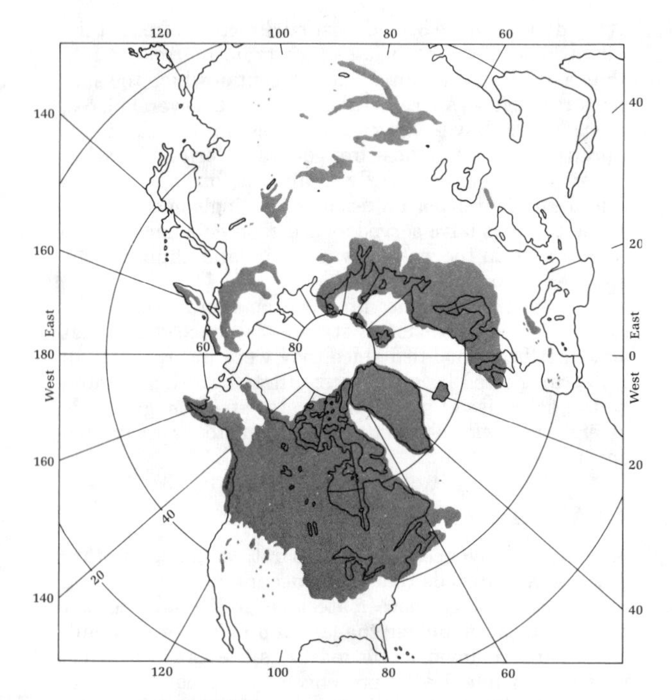

Figure 2.9
Principle areas of northern hemisphere covered by glacial ice during one or more parts of the Pleistocene Ice Age. (Adapted, by permission, from R. F. Flint, *Glacial and Quaternary Geology*, Figure 4-9; copyright © 1971 by John Wiley & Sons, Inc.)

changed and stones that have not been changed during glacial transport. Those that have been changed have snubbed corners or one or more of their surfaces abraded to nearly flat facets, most of which exhibit several nearly parallel striae (see Figure 2.4A). Both the facets and striae were formed as the stores were dragged over bedrock.

Many glacially striated stones are made up of relatively soft, easily scratched rocks such as limestone or dolostone; some, however, are made up of harder rocks such as granite. It should be kept in mind that such glacially modified stones generally make up only a relatively small percentage of all of the stones carried by glacial ice. This is true because even though a large part of a glacier's load is incorporated into it and carried near its base, only a small percentage of the fragments are actually dragged along over the underlying bedrock.

Most deposits of stratified drift are of glacio-fluvial origin—that is, they are made up of glacially transported materials that were subsequently carried and deposited by meltwater streams. Such streams flow and deposit their loads on top of, within, or near the borders of glaciers. Eskers and outwash deposits are examples of the results of such processes. Stones shaped as the result of glacio-fluvial activities are in essence stream-eroded and look like it. This, of course, means that they are indistinguishable from other stream-eroded stones. Therefore, the location from which a stone has been collected must be known if it is to be classified as glacio-fluvial versus alluvial *per se*.

As shown in Figure 2.10, several different types of glacial and glacial-margin deposits may be distinguished. Each of those illustrated may also be characterized, on the basis of its typical makeup, as follows:

Moraine chiefly, if not wholly, till
Drumlin till or stratified drift
Esker disrupted or irregularly stratified sand and gravel
Outwash well-sorted and stratified sand and gravel

Two other aspects of glacial deposition warrant mention: (1) Glacial deposits, including stones, may occur on the tops of hills as well as in valleys; and (2) glacial ice is the only natural agent that has essentially no limits as to the size of the fragments it can move. The glacial erratics (see below) on Moraine Dome in Yosemite National Park are exemplary of the former; the large quartzite boulder reported by R.L. Rutherford of the University of Alberta is an especially noteworthy example of the latter. That boulder, which occurs in southwestern Alberta, measures about 48 × 16.5 × 7.5 m (~ 160 × 55 × 25 ft), weighs about 16,500 tonnes (~18,200 tons), and

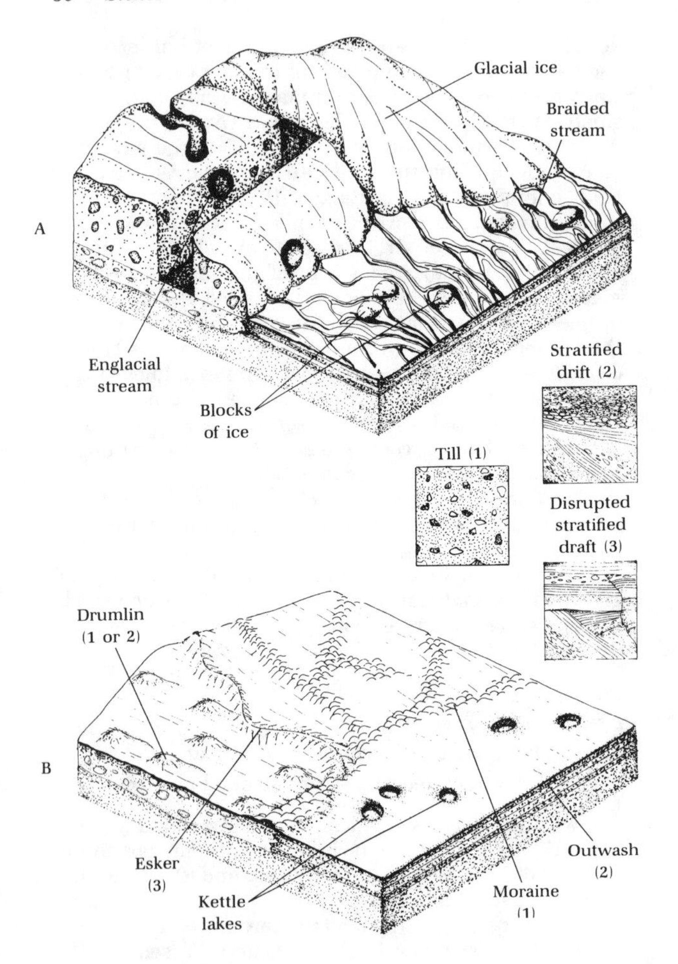

Figure 2.10
Formation of glacial and glacio-fluvial deposits. *A*: relationships during glaciation; *B*: appearance after removal of glacial ice. Square boxes to the right are cross sections indicating general characteristics of diverse deposits.

is thought to have been glacially transported some 80 to 160 km (~50 to 100 mi).

Glacially transported stones may be found within a fairly short distance of just about any place that has undergone glaciation. The best collecting is in sand and gravel pits (typically in eskers, kames, or outwash deposits), along stone fence lines or from piles within cleared fields (generally moraine or drumlin materials), and in roadcuts (see Figures 2.11 and 2.12).

Boulders transported over relatively long distances by glaciers are frequently called *glacial erratics*. Some glacially transported stones have such distinct appearances that a knowledgeable geologist can pinpoint the area from which they came. As a consequence, such erratics are sometimes referred to as *indicators*. An especially well-known example is the Lorain Jasper Conglomerate (commonly termed puddingstone) that occurs as bedrock near Sault Sainte Marie, Ontario, and has been found as stones in glacial deposits in Michigan, southern Ontario, Missouri, Iowa, Illinois, Indiana, and Ohio.

In many cases, the erratics are distributed so that they clearly point out the bedrock sources from which they came. The distributions, typically roughly linear or covering more or less triangular areas, are usually referred

Figure 2.11
Glacial boulders and cobbles are commonly collected along fence lines.

Figure 2.12
Glacio-fluvial stones are present in most sand and gravel pits in glaciated regions.

to as *boulder trains*. Each train tends to extend directly out from the bedrock area from which the stones were removed in the direction the glacier moved. Several of these boulder trains have been identified. Some of the best studied and recorded ones are in the New England states (Figure 2.13). A less well-known group is made up of porphyry boulders and cobbles that trail generally southward from each side of the laccolithic mountains in north central Montana.

When collecting glacial stones in fields or pits, two precautions are worthy of note: (1) Many fine specimens, especially along fence lines and in farm piles, have their characteristics obscured by lichen or moss—see suggestions for removing these masking growths in Chapter 5; and (2) if you are not willing to lick dry stones, as many collectors do, to see how they will appear when polished or otherwise treated to give specular reflections, carry water or some other colorless, chemically inert fluid for wetting them.

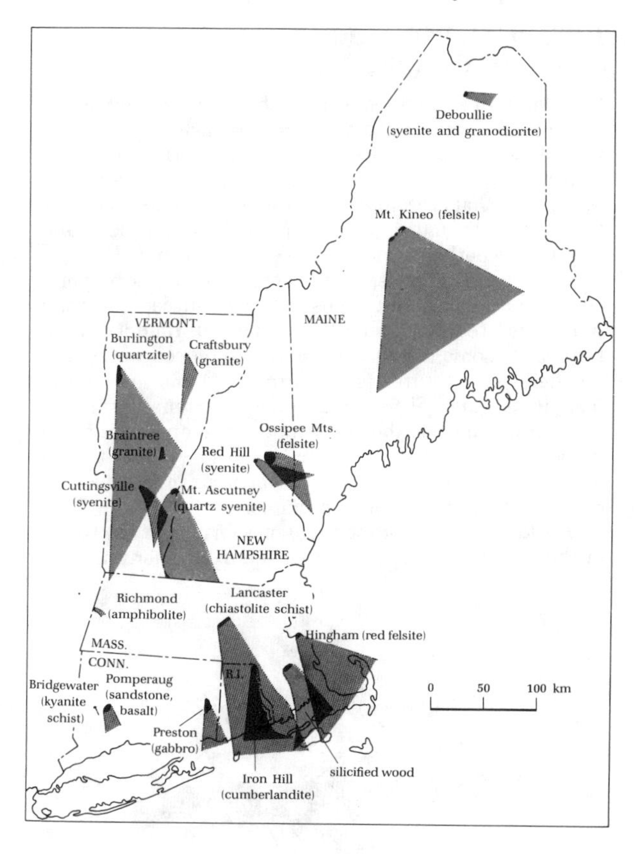

Figure 2.13
Boulder trains, many of which are dispersal areas rather than true trains, in New England. (Adapted, by permission, from R. F. Flint, *Glacial and Quaternary Geology*, Figure 7-19; copyright © 1971 by John Wiley & Sons, Inc.)

RESIDUAL BOULDERS AND LAG GRAVELS

Boulder-shaped masses of rock that have remained essentially in place since they were detached from their parent bedrock as a consequence of weathering are called *residual boulders*, or sometimes *boulders of decomposition*. Rocks that are jointed (separated along natural planar fractures) are more susceptible to both physical and chemical weathering than rocks that are not. Because corners are more susceptible to attack by weathering processes than edges and edges are more susceptible than planes, jointed rocks tend to become well rounded. Consequently, some residual boulders exhibit concentric, onionshell-like fractures (see Figure 2.14) and are said to have undergone *spheroidal weathering*. Stones formed as a result of such weathering tend to be composed of fairly homogeneous rocks; for example, basalt or fine- to medium-grained granite. Most of them occur in areas where chemical weathering predominates.

Residual concentrations of stones formed as the result of the removal of the previously admixed finer particles

Figure 2.14
Residual boulders with onionshell-like forms are fairly common in areas undergoing predominantly chemical weathering.

are generally termed *lag gravels*. Most, if not all, of the stones in such accumulations are made up of rock materials that are rather resistant under the existing climatic conditions. Thus, lag gravels may be said to form because of the different resistances to weathering or erosion offered by different materials; for example, one stone may remain more or less intact while others break down, either chemically or physically, and become susceptible to removal. The removal of the less resistant materials may also be either physical or chemical; for example, small particles may be picked up and blown away by the wind, whereas dissolved mineral matter may be carried away by percolating water. It is because of this that, in relatively warm and humid regions undergoing predominantly chemical weathering, lag gravels consist of such chemically stable minerals and rocks as quartz and chert.

Both residual boulders and lag gravels have been given all sorts of names because of their shapes, their relationships to the underlying material, their place or mode of formation, or some other characteristic feature. Examples of some of the names that have been given are: Dinosaur Rock, Elephant Rock, Fur Cap Stone, Mushroom Rock, Toadstool Rock, Witness Rock, and gara; Balanced Rock, Pedestal Stone, perched block or boulder, Rocking Stone, hoodoo, and logan; boulder belt, boulder field, boulder pavement and wall; cobble beach (also called "shingle"); desert mosaic and desert pavement; and pebble armor.

Of the more or less isolated stones, those that are balanced, perched, or "rocking" tend to attract the most attention. Each appears to be more or less precariously poised on its underlying rock (see Figure 2.15). A few may be rocked back and forth if a little force is applied, only to return to their original positions. Some are glacial erratics that have been deposited in their present positions; others are residual blocks that were detached from their underlying rocks when a less resistant, originally intervening rock was removed by weathering and erosion.

A few of the specially designated accumulations of stones also warrant further comments. *Boulder fields* and *boulder pavements*, also called *felsenmeer* (German for "sea of rock"), consist of angular rubble formed as the result of diverse processes such as by jointing *in situ* and frost heaving. *Cobble beaches*, from which the originally

Figure 2.15
Balancing rock in the Garden of the Gods, near Colorado Springs, Colorado.

admixed sand has been winnowed out, are often referred to as "shingle" because of the *en échelon* arrangement of the stones that gives them an appearance roughly resembling shingles on a roof. *Desert mosaics, desert pavements*, and *pebble armors* are made up of closely spaced, commonly touching or even overlapping, stones that occur in dry regions where the wind has removed the formerly admixed sand and silt. Some stones in these desert deposits are likely to be polished ventifacts (see below) or coated with a lustrous iron oxide-rich stain widely referred to as "desert varnish."

CONCENTRATIONS IN PATTERNED GROUND

More or less regularly shaped areas (e.g., stripes, nets, polygons, and circles) that are outlined by concentrations of stones occur in several areas where permafrost zones are or have been present beneath an unconsolidated rock-material overburden that has been alternately frozen and frost free. From above, circular concentrations look like doughnut-shaped groups of pebbles and cobbles with their "holes" made up of finer debris (see Figure 2.16).

These "doughnuts of the gods" were first described in the geological literature on the basis of occurrences in the Lapland region of northern Scandinavia and on Spitsbergen (Svalbard). Most of them range between 10 and 15 feet in diameter. They appear to have been formed as the result of a sorting that was promoted by expansion due to freezing of the water contained in the overburden, followed by a flowing outward from central areas in re-

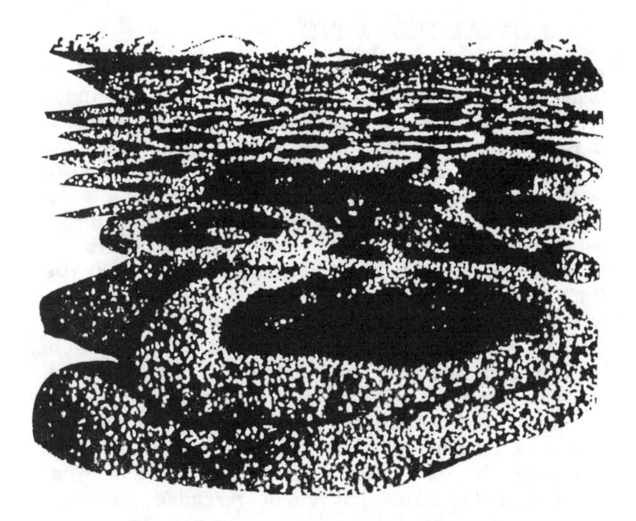

Figure 2.16
"Doughnuts of the gods," concentrations of stones appearing as circles of patterned ground. (Sketch drawn from a photograph "from the H. S. Center for Educational Resources, University of Washington, Seattle" published in *Discover*, April 1988.)

sponse to moisture differences that developed while the ice was thawing.

These features are relatively common in arctic desert and tundra regions such as those in northern North America (including the Arctic Archipelago of Canada), land areas around the Greenland ice sheet, northern Eurasia, and areas of ice-free Antarctica; they also occur in a few high mountain areas in other parts of the world. In addition, a few have been found in such places as southern Canada, New York and New England, the Appalachian highlands as far south as northern Tennessee and North Carolina, and throughout much of central Europe, where the required conditions apparently existed during the Ice Age. Wherever, stones in these concentrations are generally indistinguishable from other stones within the other overburden of the same general areas.

COLLUVIAL DEPOSITS

Rock material that has moved in response to gravity alone is generally said to have undergone *mass-wasting* and is referred to as *colluvium*. Colluvial deposits are common in many areas with steep or relatively steep topography. There are two main kinds of deposits: deposits of loose bedrock fragments that have broken off and fallen to the foot of the source cliff, and deposits formed when unconsolidated overburden merely creeps or slides downslope. The former are generally termed *talus* or *scree* (see Figure 2.17). Typically the stones in talus heaps are angular, but a few may show signs of chemical weathering, including rounding, that has taken place since they were deposited. On the other hand, the stones involved in creep are commonly rounded, in most cases in response to chemical processes enhanced by percolating groundwater solutions. On some hillsides, generally near their bases, stones that have crept downslope have become somewhat concentrated in strips that roughly resemble terraces.

COLLAPSE RUBBLE

A somewhat unusual kind of stone that has been formed and moved in response to gravity occurs locally where cavern roofs have collapsed. The individual fragments may range from rather large to fairly small and from

Figure 2.17
Talus. This deposit is kept active by the stream, which is continually removing stones from the toe area of the talus slope.

angular to subrounded. They may be found in some caves and in the bottoms of some sinkholes.

Although the stones of these deposits tend to be predominantly limestone or dolostone fragments, they also may include other sedimentary rocks that comprised rock units that were interstratified with the cavern roof rocks or stones of just about any composition that were within the overburden that was on top of the roof rocks. In addition, some of these deposits include recognizable fragments of or even complete speleothems (cave deposits of chemical origin) such as stalactites and stalagmites.

SLIDING STONES

Stones that occur at the ends of what appear to be skid tracks on playa lake beds are frequently referred to as *sliding stones* or *skid boulders*. (*Playas* are the shallow, intermittently dry lakes found in closed depressions in semiarid and arid regions.) One such playa in southeastern California that is well known because of these stones is called Racetrack Playa (see Figure 2.18). The stones, which come from the nearby upland, range in weight from about 200 g (~ ½ lb) to as much as 25 kg (~ 55 lb). The tracks indicate that the movements of the

Figure 2.18
Sliding stones. Sketch is of area of Racetrack Playa, California. (Sketch drawn from photographs in R. P. Sharp and D. L. Carey, 1976, *Geol. Soc. America Bull.*, v. 86, p. 1704–1717.)

stones have been rather erratic. It is now thought that most of the movement takes place during the winter months and is dependent upon pushing by the wind over slippery clay or iced-over surfaces. Aerodynamic considerations indicate that lee side vorticity and drag may also contribute to the movements.

RAFTED STONES

Stones that have been carried over water by plants or by ice are said to have been *rafted.* Examples are stones carried by driftwood (see Figure 2.19), stones transported on mats of swamp growth ("floating islands"), and stones carried on or in floating ice (including icebergs; see Figure 2.20). Stones carried by driftwood and by floating ice are fairly common. Those carried by floating islands must be rare, if indeed they exist. Many rafted stones have been found hundreds of miles from their bedrock source areas. Some stones may be confidently inferred to have been rafted. Two examples are glacially striated boulders that occur well away from any glaciated area and large stones that have obviously been dropped into lacustrine silt or clay. Perhaps the best example of the former are the iceberg-transported stones that have been found in low-latitude ocean-bottom sediments.

Figure 2.19
Stones are sometimes enmeshed in tree roots and subsequently transported as the roots become driftwood.

ICE-PUSHED STONES

Here and there, at or near water level, along the shores of some of the water bodies that are frozen each winter, there are lines of compactly arranged boulders. Good examples have been reported as occurring sporadically along the coasts of Labrador, the Saint Lawrence River, the Great Lakes, and even along a number of the smaller streams and lakes that are located at relatively high lat-

Figure 2.20
Stones are carried by most icebergs, which are calved-off pieces of glaciers that flow into ocean water. (Sketch is from the 1856 report "Arctic Explorations: The Second Grinnell Expedition in Search of Sir John Franklin, 1853, '54, '55" by E. K. Kane.)

itudes or high altitudes. Most of these boulders are thought to have been moved to their present positions as the result of expansion of or pushing by ice. In any case, the boulders are too large to have been moved in response to simple wave or current actions. Some groupings of ice-pushed boulders look like crude walls. As a result, they have been called *boulder barricades*. Those along high-latitude seacoasts are generally most obvious between low and half tide. Wall Lake in Sac County, Iowa is said to have been so named because of the presence of such boulders around much of its perimeter.

ANIMAL-TRANSPORTED STONES

Humans have probably transported many more stones than all other animals combined. In fact, stones transported by animals other than humans are rare. On the other hand, one can only guess at how many stones occur in odd places because they have been dropped there by a human. At one time, for example, a street in the dock district of New Orleans had a stone pavement made up of stones carried in as ballast by ships from all over the globe.

Perhaps the most peculiar mode of transport by animals is as gizzard stones, which are frequently called *gastroliths*. Those from reptiles, including dinosaurs, and birds have been identified from several localities. Some primitive peoples believed such stones, which are generally very highly polished, to be religious charms.

VENTIFACTS

Stones whose shapes and surface features have resulted from abrasion by windblown sand or dust particles commonly have characteristic, though diverse, shapes. No matter what their shapes, however, all stones that are so modified are called ventifacts. Most wind-fashioned stones appear to have been beach, river, or glacially transported and deposited stones before they were eroded by windblown projectiles. By close examination, some geologists can decipher the directions of the prevailing and other winds that were effective during the development of any given ventifact. Most ventifacts are found in areas relatively free from vegetation; for example, in lag gravels in corridors between sand dunes.

OTHER OCCURENCES

Some kinds of stones have not gone through the previously mentioned rock-to-stone transition as the result of terrestrial weathering and erosion. Instead, they have come to the earth from outer space; they have been formed from molten rock material that cooled and solidified while it was flying through the air; or they have been precipitated from solutions within solid rock, within unconsolidated sediment or overburden, or within water. Examples, listed respectively, are meteorities, tektites, volcanic bombs, Pele's hair, and diverse concretions.

Meteorites

Meteorites are stones from outer space. They are of three main kinds: those made up almost entirely of metallic iron-nickel alloys ("iron meteorites"), those made up almost wholly of nonmetallic minerals ("stony meteorites"), and a transitional group ("stony-iron meteorites").

As recently as the sixteenth century, these stones were thought to have quite different origins. For example, some people thought them to be terrestrial rock materials that were swirled up by powerful hurricanes and then dropped; others suggested that they were stones ejected from volcanoes on the moon; and several continued to subscribe to the ancient idea that they were gifts from the gods (and thus should be worshipped).

Although meteorites may have all sorts of irregular shapes, many are roughly conical and have grooves and pits on their surfaces that were apparently formed as the result of inflight aerodynamic abrasion (see Figure 2.21). Statistical analyses of meteorites seen falling versus those already recovered indicate that there are many meteorites yet to be found, especially those of the stony category. Meteorites are apt to be found just about anywhere.

Tektites

Tektites are masses of yellowish to dark green, dark brown, or black glass that range from microscopic to about 15

Figure 2.21
A: Meteorites. These stones from outer space commonly exhibit aerodynamic abrasion, as shown in *B*. Polished and etched surfaces of iron meteorites may have patterns like that indicated in *C*.

cm (~ 6 inches) in greatest dimension. Their shapes are rather varied and often described on the basis of some obvious resemblance to a relatively well-known object. A few rather common examples are bean-, bowl-, button-, cudgel-, disc-, dumbbell-, gherkin-, and peanutlike; and lens-, pear-, teardrop-, and irregular-shaped. Also because of their shapes, tektites have been called such ridiculous things as "Blackfellows' buttons" by some of the early white settlers in Australia and "emu eyes" by some of the aborigines of the same general area. The surfaces of tektites range from smooth to rough, commonly exhibiting numerous grooves and pits (see Figure 2.22).

Both the shapes and surface features of tektites indicate formation under the influence of aerodynamic forces. Although they were once referred to, both scientifically and romantically, as "moon splash," today it is generally agreed that they have been formed when hypervelocity meteorites, asteroids, or comets have hit the earth and caused shock melting and subsequent splattering through the air of the terrestrial rock materials that were so melted.

In all known cases, tektites occur in strewn groups consisting of literally thousands, if not millions, of individual stones. The tektites of the Bohemia district of Czechoslovakia (called *moldavites*) and from the northwest-southeast-tending belt that extends from Indochina, through Java and the Philippines, and across Australia

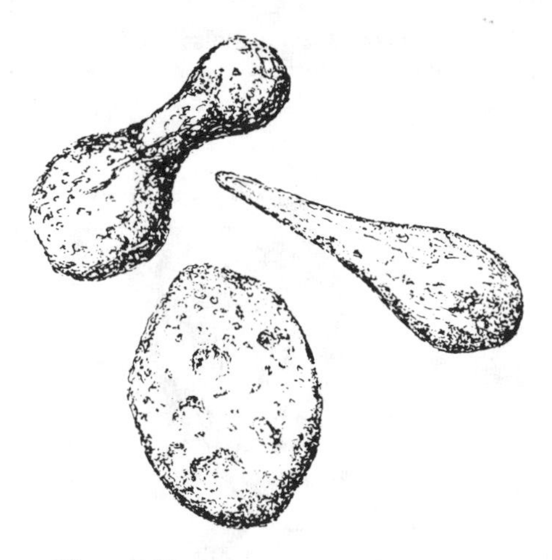

Figure 2.22
Tektites tend to have shapes like these, no matter what their size.

(called *indochinites, philippinites, australites*, and so forth) are exemplary.

Volcanic Bombs

Volcanic bombs are formed from lava that is ejected in a molten or highly viscous state and then is consolidated —either completely or in part—while in flight. These "bombs" may range in size from a few millimeters to several centimeters in greatest dimension. They are readily recognized because of their turniplike shapes and twisted ends (Figure 2.23). In addition, nearly all of them are brown in color and many have a characteristic network of surficial cracks that are often compared to toasted breadcrust. Many are relatively light in weight because their interiors are scoraceous (full of holes). Most are associated with andesitic or basaltic composition vulcanism such as that of south-central Idaho and that in the area around Amboy Crater in southern California.

Pele's Hair and Pele's Tears

Pele's hair and *Pele's tears* are also products of volcanic eruptions. (Pele was the Polynesian goddess of the vol-

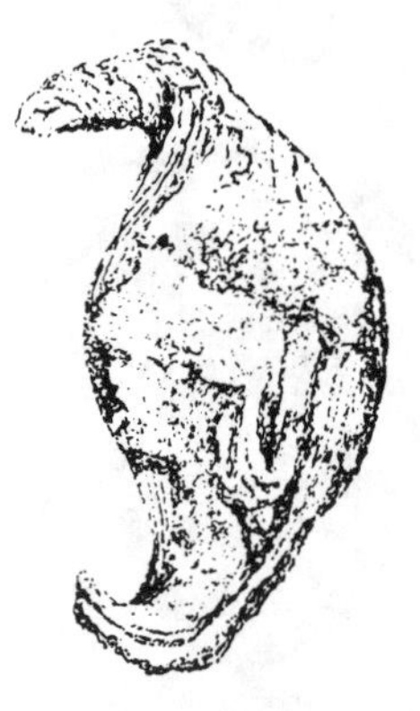

Figure 2.23
Volcanic bombs tend to have turniplike shapes with whorled ends.

cano Kilauea.) The hair consists of threadlike fibers of basaltic glass that are formed when bubbles of lava burst and the attenuated lava is pulled out into the fibers. Each fiber is typically less than a half millimeter in diameter and ranges up to a meter or two in length. Most frequently, many hairs are found as matted masses that appear to have been blown and collected by the wind. The tears, which commonly occur with the hair, are solidified drops of lava. Some, in fact, have trailing ends that are indistinguishable from the hair. Both are typically a golden brown color, and do indeed look very much like human hair. Unlike human hair, however, the individual strands of Pele's hair are brittle and thus break very easily. The similarity in appearance, however, once led to the confiscation by a U.S. Customs officer of some Pele's hair that I had collected. Although the Pele's hair was clearly marked as "rock material," it was removed and in its stead there was an explanation to the effect that "organic materials cannot be shipped into the United States. . . ."

Cavity Fillings

In the broad sense, the designation cavity fillings as applied to rocks includes several things such as joint and fissure fillings (veins), some of which constitute ore deposits. Only those that would likely result in the formation of ellipsoidal or spheroidal masses that might tend to weather out and become loose stones or stonelike masses are dealt with here.

Vesicles, sometimes called vacuoles, are cavities produced by expanding gas bubbles formed within magmas during their consolidation. They are particularly common in basaltic rocks (see Table 3.1). If subsequently filled, vesicles are referred to as *amygdules*. In some cases, the mineral constituents of amygdules are more resistant to weathering than the surrounding rock, and thus weather out and become stones. The stones may occur in residuum or be transported and become, for example, stones in streams. The agates and several other beach stones that once filled cavities in the ancient volcanic rocks of Isle Royale and some of the mainland areas around the western part of Lake Superior are good examples. Included are the pebbles that exhibit thomsonite "eyes," chatoyant and patterned chlorastrolite (i.e., the pumpellyite "greenstone" that is Michigan's official state gem), and the diversely colored datolite masses—all of which are eagerly sought by collectors and widely used by lapidaries.

Thundereggs are ellipsoidal or spheroidal masses that have filled cavities in volcanic rocks, typically of rhyolite composition (see Table 3.1). Thundereggs were so-named for occurrences in Oregon, where Indian folklore holds that the thunder spirit of Mount Jefferson and Mount Hood threw the masses at each other during fits of anger. It was also held—at least by some—that the spirits got the eggs by robbing nests of "thunderbirds." Ranging up to a few inches in diameter, these "eggs" consist largely of silica plus feldspar border zones and chalcedony cores. Some of the cores are made up of colorfully banded or plumed agate. The official state rock of Oregon, thundereggs also occur in California, Idaho, Nevada, and New Mexico—for example, at Rockhound's State Park, which is located about 15 miles southeast of Deming. A few specimens from at least some of the localities exhibit the feature shown in the sketch, Figure 2.24. It would seem that the host rock unit was tilted during the period when the agate was being deposited to form the core; this conclusion is based on the principle that the materials of the

A

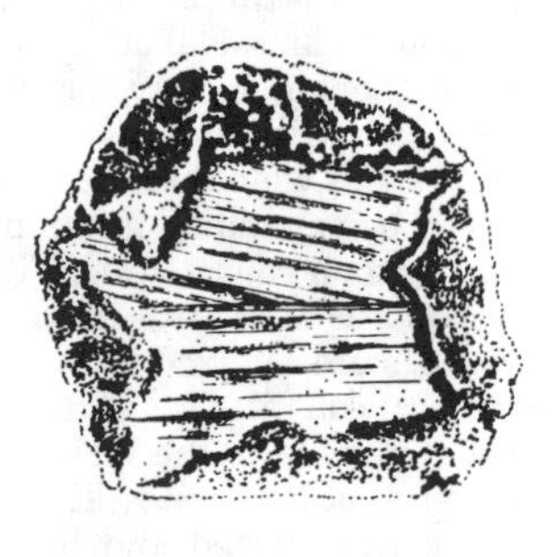

B

Figure 2.24
Thundereggs. *A*: external surface, *B*: cut and polished slice.

flat planes, which are represented by the nearly horizontal bands of the sketch, were deposited when the planes were essentially horizontal.

Vugs are relatively small cavities—typically crystal-lined—most of which occur in relatively soluble rocks such as limestones and dolostones. Some vugs become completely filled and, albeit infrequently, those fillings subsequently become weathered out and become stones.

Geodes resemble vugs but, by definition, have border zones that differ in composition from and thus are easily separable from their host rocks. Although geodes sometimes weather out from their enclosing rocks, only a few persist as stones. This is so because their crusts, typically rather thin, and their central voids tend to make them fragile and thus easily broken during transport.

Concretions

Concretions may develop in solid rock, unconsolidated sediments or overburden, or even within water. In essence, each concretion represents a mass of rock and mineral matter that has distinct boundaries with its enclosing materials. Most concretions appear to have been formed by the deposition of minerals around some nucleus—in some cases, by deposition within open space, including available pore space; in others, as the result of replacement; in still others, by some combination of these processes.

Most concretions range from a few millimeters up to a few meters across. Although many are essentially oblate spheroids, others have all sorts of remarkable shapes, some exhibiting roughly bilateral symmetry (see Figure 2.25). All tend to have their shortest axes essentially perpendicular to the stratification. The common minerals that either singly or in some combination are present within concretions in greater proportions than they are in the surrounding materials are calcite, quartz (either macro- or microcrystalline), hematite, limonite, siderite (an iron carbonate), pyrite, or marcasite (a second form —along with pyrite—of iron sulfide).

Concretions within rock sequences may lie between strata or may cut across and include the stratification of the enclosing rocks. The former relationship indicates

Figure 2.25
Concretions.

that those concretions were formed after the underlying strata were laid down but before deposition of the overlying stratum. The latter relationships, on the other hand, indicate that those concretions were formed within the enclosing unit—that is, after the unit was deposited.

As already noted, concretions have also been developed in water: in seawater, spring water, and underground pools. The manganese nodules that blanket large areas of the ocean floor and are judged to constitute large reserves of metallic manganese exemplify those formed within seawater. "Cave pearls," which are formed in cavern pools, exemplify those formed in fresh water. Cave pearl, by the way, is a truly fitting designation for these speleothems because each "pearl" is pea-sized or smaller and is typically composed of concentric shells of pure white calcite.

The clay-ironstone concretions referred to in Chapter 1 are of special interest from an economical as well as historical standpoint (see Figure 2.26). This is true because they are relatively common in some of the gravels that are used as concrete aggregate. For some purposes, such gravels must be sorted, frequently at great expense, and the concretions discarded because of the deleterious effect they have on the concrete. That is, they often cause breakage ("pop-outs") and they also may cause the development of unattractive rusty brown stains.

Perhaps the most eagerly sought concretions are those that contain well-preserved fossils. One group, especially well known in collecting circles, occurs within some approximately 300-million-year-old strata of the coal-bearing measures that occur in the vicinity of Mazon Creek,

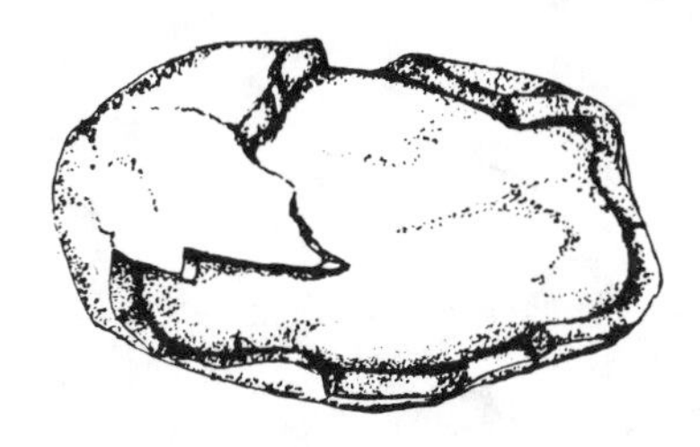

Figure 2.26
Clay-ironstone concretions commonly consist of a hard iron-rich outer layer surrounding relatively soft clay or mud.

northeastern Illinois. The fossils there, which appear to have served as the nuclei around which the concretions developed, include replacements of such delicate organic remains as fern leaves and insect wings. As might be expected, some of the best of these fossil-containing concretions have been collected from spoil heaps at some of the coal mines.

Concretions of each of the diverse origins are frequently found loose—that is, as stones. Perhaps somewhat surprisingly, this is even true of concretions that have been enclosed within rock. This reflects the fact that, in general, concretions weather out whole because they tend to resist nearly all kinds of weathering much better than their typically less resistant enclosing rocks.

Quasi-Sandstone "Stones"

In the first edition of this book, an answer was given for the often-posed question, "Do stones grow?" By implication, it said: "No, stones do not grow."

Since that time, however, I have seen and studied "stones" that have indeed grown and, so far as I know, some of them are still growing. These "stones," admittedly odd-balls, have formed as the result of subsurface growth of fungus within beach sand (see Figure 2.27). These masses occur here and there near the shores of Lake Michigan and Lake Superior. The fungus grew from a substrate of buried tree branches, permeated the sand

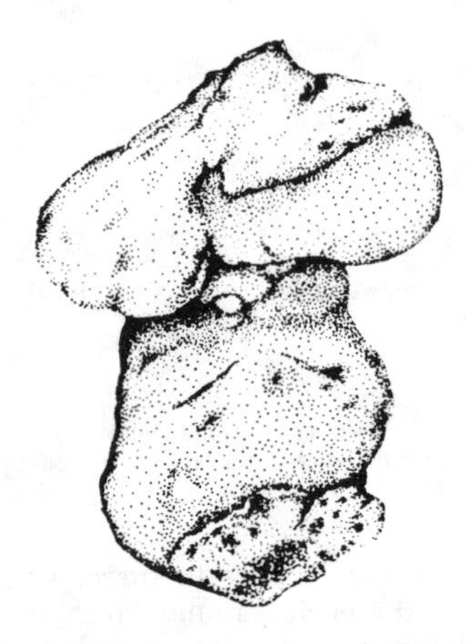

Figure 2.27
Quasi-sandstone mass with overall shape of a toadstool. This "stone" has undergone little transport subsequent to being separated from its substrate.

above it, and thus produced solid, stonelike masses. The resulting fungus-cemented quasi-sandstone masses are solid to the degree that some of them have withstood activities of the surf and are still indurated enough to be thrown through a glass window.

3

Origins of Parent Rocks

As mentioned in the preceding chapter, every stone has had at least two main steps in its history: the first establishing its mineralogical or lithological identity, and the second determining its shape and location. This chapter deals largely with the former: how rocks are formed.

The crust or outer shell of the earth is made up of rocks. Most geologists classify rocks into three main categories—igneous, sedimentary, and metamorphic—based on their origins. Like many classifications of natural substances and phenomena, however, this classification is an oversimplification; some rocks fit better along boundaries between two of the main categories than within either one. Nonetheless, as long as its limited application is kept in mind, the classification is of value, especially to anyone interested in how different rocks have been formed (see Figure 3.1).

Igneous rocks are formed when molten rock material, called *magma*, is cooled and solidified. The cooling may be rapid and result in the formation of natural glass or it may take place more slowly, permitting crystallization of minerals. Magma may be solidified beneath the surface of the earth, or it may come out onto the surface—in which case it is generally called *lava*—to be cooled and solidified there. There are several kinds of igneous rocks because magmas differ in composition and because they cool in different ways and at different rates.

Sedimentary rocks have two principal modes of or-

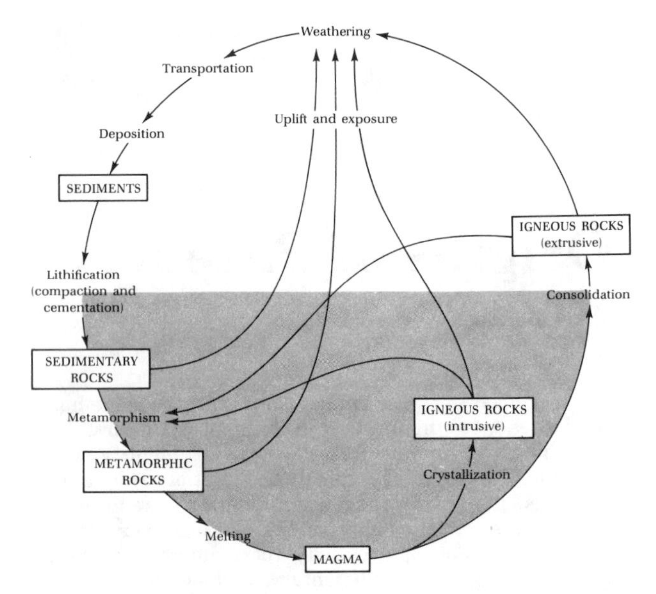

Figure 3.1
The rock cycle indicates possible interrelationships among the major kinds of rocks. Rocks and rock materials are enclosed in boxes; processes are not. Shaded and unshaded areas represent subsurface and surficial domains respectively.

igin. Some are formed by compaction or cementation of detritus, such as sand, that has been transported and deposited on the surface of the earth; others are formed by precipitation of substances from natural aqueous solutions that are on or near the surface of the earth. Most sedimentary rocks are layered. Many contain fossils.

Metamorphic rocks are formed by the transformation of preexisting rocks. Metamorphic rocks are marked by changes in arrangement of constituents, changes in mineralogical makeup, changes in overall composition, or by some combination of these changes. High temperature, high pressure, and chemically active fluids promote the changes—usually termed *metamorphism*—when rocks are subjected to geological environments notably different from those in which they were originally formed.

Other rocks, most of which combine characteristics of two of the main classes, include (1) sediments markedly changed by diagenesis, (2) pyroclastic rocks, and (3) migmatites. Also included are (4) pegmatites that are not of igneous origin and (5) different kinds of coal and other solid, natural organic materials. Sediments that have undergone marked diagenetic changes fit along the sedimentary-metamorphic classification boundary; pyroclastic rocks belong on the igneous-sedimentary boundary; many migmatites straddle the metamorphic-igneous boundary. Nonigneous pegmatites are deposited from high-temperature, low-viscosity fluids of various origins. The diverse coals and related materials consist largely of organic compounds rather than of minerals or glass; consequently, some geologists do not even consider them to be rocks.

Most rocks have one or more features that permit their ready identification (see Appendix C). Some rocks and the stones derived from them exhibit features that permit deciphering rather complex histories (see Figure 3.2).

IGNEOUS ROCKS

As shown in Table 3.1 (page 58), the name of a crystalline igneous rock depends upon the kinds, proportions, and general grain size of the main constituent minerals. Many petrologists (the geologists who are interested chiefly in rocks) consider the component minerals to be of four main kinds. *Specific minerals* are those required by the definition of the given rock. *Varietal minerals* are those commonly present in noteworthy quantities but not required to name the rock; their names are generally used as adjectives in complete rock designations. *Accessory minerals* are those present in very small amounts; generally they are not alluded to in a rock's designation. *Alteration products* may or may not occur; if present, they may or may not be referred to in the overall name. When they are mentioned, the suffix *-ized* is frequently used to call attention to the fact that these minerals are believed to have formed as the result of something that happened to the rock either during the late stages of or after its consolidation from magma.

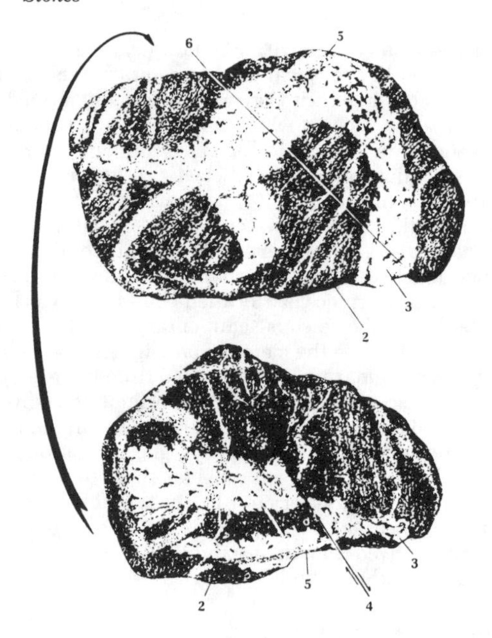

Figure 3.2

Some stones have had complex histories that can be deciphered: This boulder was collected from a glacio-fluvial deposit located about 11 km (~ 7 miles) west of Mount Pleasant, Michigan, nearly 160 km (~ 100 miles) southeast of the nearest area with metamorphic rock bedrock. The steps outlined below constitute a probable history—"probable" because steps 2, 3, and 4 could have taken place in a different order. Step 2 could have followed step 6; and no-longer-recognized, additional steps could have occurred.

1. *Sedimentation and lithification.* Sediments, at least in part derived from volcanic eruptions, were deposited and subsequently converted into sedimentary rock called *graywacke*.
2. *Deep burial and metamorphism.* Later, while buried rather deeply, the graywacke was submitted to conditions of high temperature and high pressure and converted into the metamorphic rock called *amphibolite*.
3. *Magmatism.* Igneous magma was injected into the rock and, upon cooling, consolidated to form the granitic dikelet.
4. *Deformation.* The rock, including the dikelet, was affected by differential stresses and as a consequence was faulted.
5. *Vein Formation.* Silica fluids penetrated the faulted granitic

Kind of Mineral	*Mineral*	*Percentage of Volume*
Specific	Quartz	18.0
	Alkali feldspar	8.0
	Oligoclase (a plagioclase feldspar)	45.0
Varietal	Biotite	10.0
	Hornblende	16.0
Accessory	Apatite	0.4
	Zircon	0.3
	Others (magnetite, etc.)	0.3
Alteration product	Chlorite (as alteration product of some of the biotite)	2.0
		100.0

As noted in the tabulation: (1) The specific minerals are quartz and the two feldspars. The plagioclase content far exceeds the alkali-feldspar content, so the rock is a granodiorite (see Table 3.1). (2) The varietal minerals are biotite and hornblende; thus, the rock is a biotite hornblende granodiorite. (3) The accessories are apatite, zircon, sphene, magnetite, and so forth; their presence does not change the rock designation. (4) Alteration is indicated by the chlorite alteration product of some of the

---

dikelet and the surrounding amphibolite to form veins of the mineral quartz.

6. *Vein formation*. At a later time, silica fluids again formed quartz veins—these cut across all of the other features including the earlier formed (#5) veins.
7. *Erosion*. The rock was exposed at the surface as a result of erosional removal of overlying rocks.
8. *Transportation*. A piece of the rock was loosened and transported to the glacio-fluvial deposit where it was collected. The transport was glacial and glacio-fluvial and perhaps also included colluvial and fluvial processes prior to the effective glaciation.

Table 3.1 Igneous Rock Chart

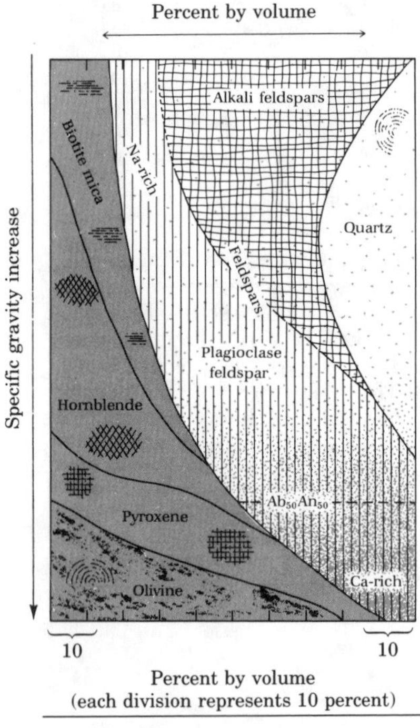

Percent by volume

Specific gravity increase

Alkali feldspars
Biotite mica
Na-rich
Quartz
Feldspars
Plagioclase feldspar
Hornblende
$Ab_{50}An_{50}$
Pyroxene
Olivine
Ca-rich

10 10

Percent by volume
(each division represents 10 percent)

Phanerites	*Aphanitic Equivalents*
Syenite	Trachyte
Granite	Rhyolite
Granodiorite	Dacite
Diorite	Andesite
Gabbro	Basalt
Peridotite	

SOURCE: After R. V. Dietrich, *Geology and Virginia*, copyright © 1970, University Press of Virginia.
Note: Diagram shows the relative proportions of the chief mineral components of common igneous rocks. Adjacent rocks listed in the vertical columns grade into one another. The $Ab_{50}An_{50}$ compositional division in the plagioclase field, which serves to distinguish diorites from gabbros, cannot be determined megascopically.

biotite; therefore, the rock could be designated as a partially chloritized, biotite hornblende granodiorite.

As already mentioned, the grain size of an igneous rock tends to reflect the speed with which its magma cooled and consolidated. Generally speaking, the more rapid the cooling the smaller the grain size. The cooling rate, of course, depends upon the environment or environments in which a magma is consolidated. Therefore, the grain size of an igneous rock—considered in conjunction with its composition—provides an important clue as to how and where the rock was formed.

As also mentioned previously, magma may solidify beneath, as well as on top of, the earth's surface. Igneous rocks and masses formed below the surface are generally termed **intrusive**, whereas volcanic rocks and masses formed on the surface are generally termed **extrusive**. Strictly speaking, however, there are two main types of igneous masses formed below the surface: those formed from magma that was consolidated essentially where the magma was generated, and those formed from magma that solidified after the magma had moved well away from the place it was generated. The latter make up the *injected* igneous masses. Most intrusive masses have been given names such as *batholith, laccolith, dike*, and *sill*, names that are based on the overall shapes of the masses or their relationships with the surrounding country rocks (i.e., the rocks into which magma has been intruded) (see Figure 3.3).

Most masses of magma that have cooled relatively slowly (for example, at fairly great depths within the earth's crust) have their specific and varietal mineral grains large enough to be distinguished by the naked eye or by the eye aided by only a 10× handlens. Such rocks may be described as **phaneritic** (see Figure 3.4). On the other hand, rocks that are formed as the result of very rapid cooling of a magma (for example, those formed from lava flows or in zones directly adjacent to relatively cool country rocks) typically have a grain size so small that they appear stony in hand specimens or else they are glassy. Stony-appearing rocks are termed **aphanitic**; their constituents can only be seen with the aid of a high-powered microscope. Glasses are noncrystalline; their compositions have to be determined chemically. On an

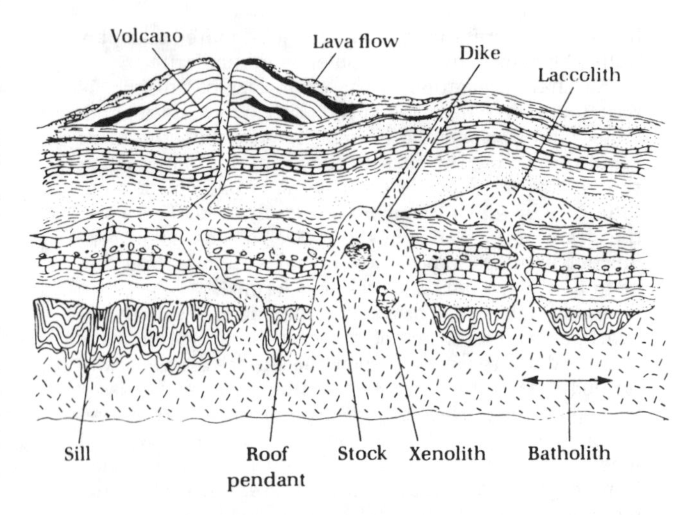

Figure 3.3
Some of the more common igneous masses. (Modified after R. G. Schmidt and H. R. Shaw, U. S. Geological Survey)

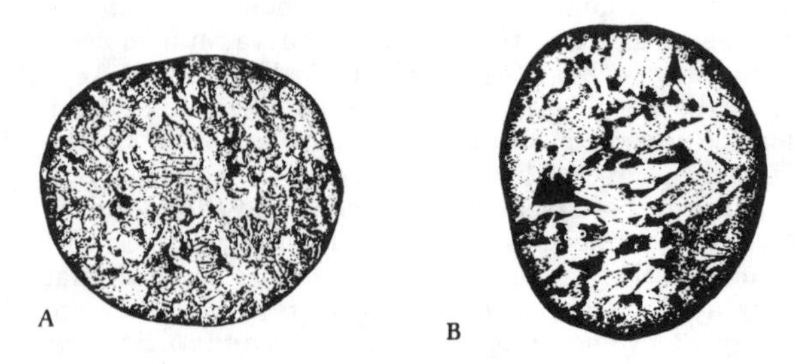

Figure 3.4
Two typical phanerites. *A*: Granite. *B*: Syenite. (Figure 1.3 also shows a phanerite.)

empirical basis, however, we know that most natural glasses may be safely called *obsidians*. The exceptions, which are directly associated with dark rocks such as basalt, are called *tachylytes*.

Two other fairly common grain-size designations that are applied widely to igneous rocks are **pegmatitic**, which describes extremely coarse-grained rocks, some of which contain individual crystals up to a few meters long; and **porphyritic**, which describes rocks that contain grains distinctly larger than the vast majority of the constituent grains (see Figure 3.5). Some pegmatitic rocks may not be of igneous origin and thus are treated in the section of this chapter entitled Other Rocks. In porphyritic rocks, the distinctly larger grains are called *phenocrysts*, whereas the fine-grained matrix that surrounds the phenocrysts is generally termed *groundmass*. Many porphyritic rocks appear to have been formed when early cooling of their magma was slow, so the relatively large phenocrysts crystallized, and later cooling was much more rapid, so the remaining magma cooled to form the finer grained groundmass.

Many fine-grained gabbros (frequently called *diabase* or *traprock*) and basalts are **vesicular**—that is, they are filled with holes from which gas escaped. In some of these rocks, the vesicles have been filled with minerals and the rock is then said to be **amygdaloidal** (see Figure 3.6). Although they resemble each other, most amygdaloidal and porphyritic rocks are readily distinguished because amygdules tend to be spheroidal, whereas phenocrysts tend to have crystal outlines. Many stones that consist wholly or largely of certain individual minerals have been derived from amygdules and other cavity fillings in an-

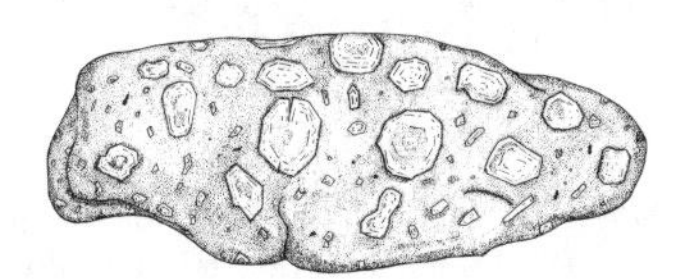

Figure 3.5
Two phorphyries.

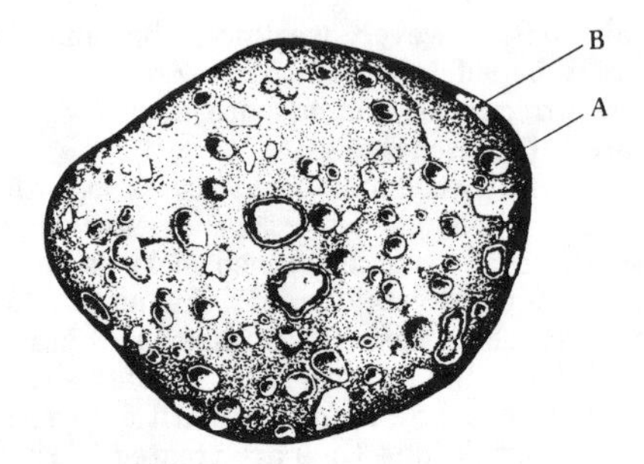

Figure 3.6
Amygdaloidal porphyritic basalt. Amygdules (*A*) are typically spheroidal and thus rather easily distinguished from phenocrysts (*B*), most of which exhibit at least some angularity.

cient lavas. Examples are stones constituted by agate, native copper, datolite, and thomsonite.

Rocks composed of quartz and potassium feldspar that appear intergrown in such a way that the quartz grains have shapes that on some surfaces are suggestive of cuneiform writing are called *graphic granite* (see Figure 3.7). Fine-grained, sugary-textured rocks composed of quartz and both an alkali- and a plagioclase-feldspar are

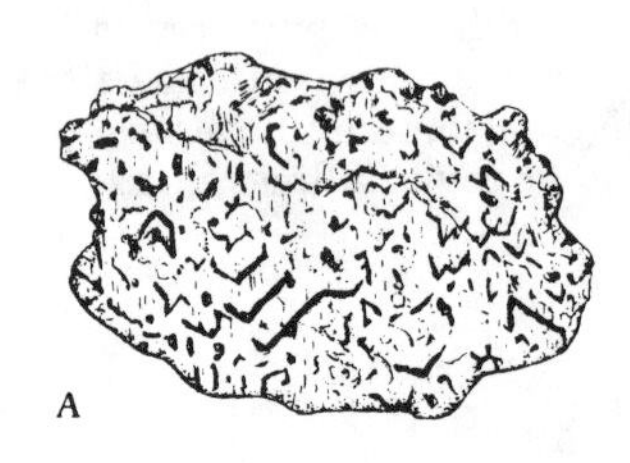

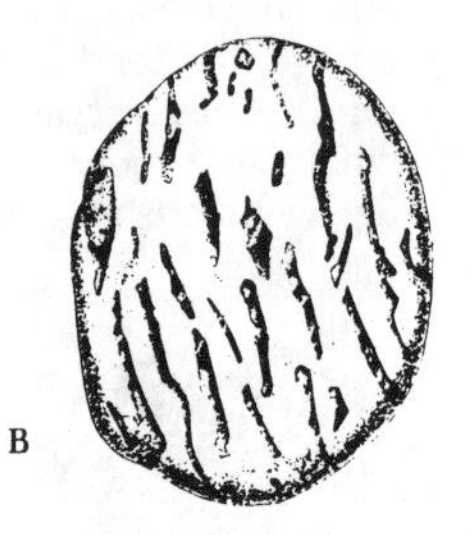

Figure 3.7
Two views of graphic granite. *A*: Cross section showing the relationships that led to the name—that is, what is seen essentially perpendicular to the quartz rods (the darker of the two components). *B*: Stone viewed nearly perpendicular to the cross section shown in *A*.

called *aplite*. Graphic granite and aplite are commonly associated with each other and with granitic pegmatite.

SEDIMENTARY ROCKS

The characteristic layering of sedimentary rocks (see Figure 3.8) is usually termed *stratification*. In fact, the geologists who study the diverse aspects of, and the interrelationships among, sedimentary rocks are called *stratigraphers*.

Sedimentary rocks may be classified into two main categories on the basis of how they were formed: **Detrital** sedimentary rocks (see Figures 3.9 and 3.10) consist largely of rock and mineral fragments that have been transported

Figure 3.8
Stratification (layering) in sedimentary rocks is a manifestation of such things as different colors, sizes, shapes, sorting, or compositions of the constituents of the different layers. The four sketches show different stones made up of well-stratified sedimentary rocks.

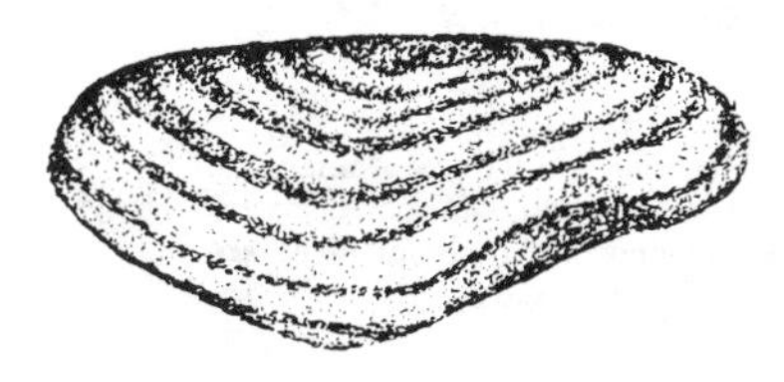

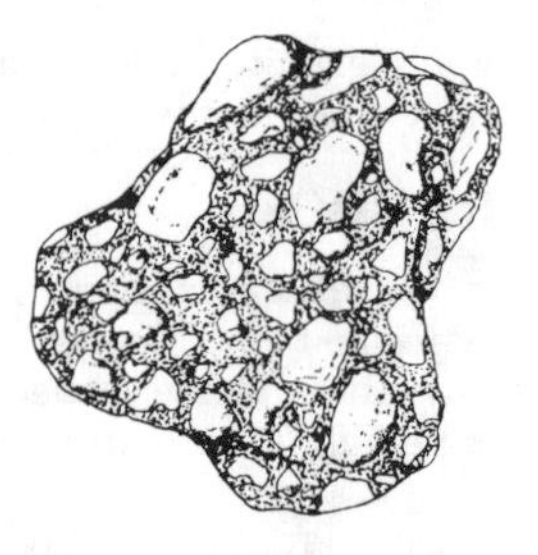

Figure 3.9
Conglomerate is the name given to gravel that has been cemented into solid rock. Many conglomerate stones are knobby because the sandy portions tend to erode more readily than the pebbles.

and deposited as the result of gravitational forces or by water, ice, or wind on the surface of the earth; **nondetrital** sedimentary rocks consist wholly or largely of mineral grains that have been chemically or biochemically precipitated from aqueous solutions on or near the surface of the earth. The terms *(non)clastic* and *(non)-fragmental* are used by a few geologists instead of *(non)detrital*.

Most detrital rocks are named on the basis of the size of the grains that make up 50 percent or more of the rock. A few names also depend upon grain shape (see Table 3.2) or mineral composition (see below). The coarser de-

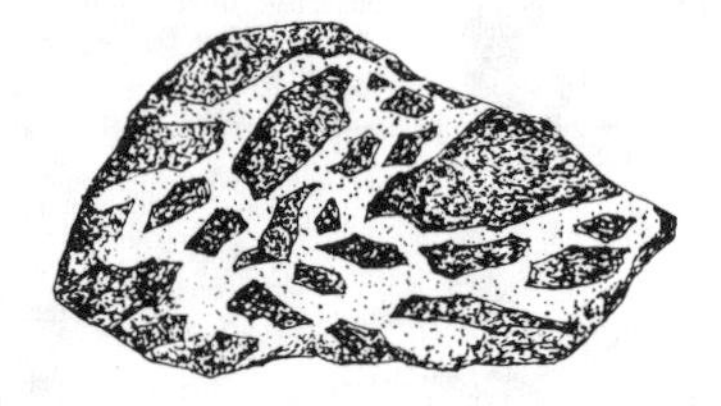

Figure 3.10
Sedimentary breccias are like conglomerates, except that the larger fragments of breccias are angular whereas those of conglomerates are rounded.

Table 3.2 Detrital Sedimentary Rocks

Diameter of Fragments (mm)	*Loose Aggregates*	*Consolidated Aggregates*	*Remarks*
> 2	Gravel	Conglomerate	Rounded fragments
	Rubble	Breccia*	Angular fragments
1/16–2	Sand	Sandstone	
1/256–1/16	Silt	Siltstone	
< 1/256	Clay ("mud")	Claystone	
		mudstone	
		shale	If fissile—that is, if it will split along fairly closely spaced planes parallel to bedding laminae

*Many breccias are not of sedimentary origin.

trital sedimentary rocks represent ancient stone-bearing deposits. Frequently these conglomerates and breccias are distinguished by names that include the size of their predominant fragments (e.g., *pebble conglomerate*) or the mineral or rock name of their predominant constituent (e.g., *jasper breccias*). A few conglomerates and breccias have had their origins determined and thence have been given more definitive designations; for example, consolidated glacial deposits made up of unsorted drift have been termed *tillites*.

Most nondetrital sedimentary rocks are named on the basis of their mineralogical compositions (see Table 3.3).

Of the three major categories of rocks, only sedimentary rocks commonly contain recognizable fossils (see Figure 3.11). Richly fossiliferous sedimentary rocks are often named accordingly; for example, fossiliferous limestone, crinoidal limestone, and coral-rich limestone. Some fossils indicate the environment in which their containing rock material was deposited; for example, in marine versus non-marine water or in shallow versus deeper water. The standard method of naming fossils and a tabulation of animal and plant groups well represented in the fossil record are given in Appendix B.

A few relatively common detrital sedimentary rocks that have distinctive compositions or features have been given names that are usually used instead of, for example,

Table 3.3 Nondetrital Sedimentary Rocks

Rock Name	*Chief Mineral Constituent*
Limestone*	Calcite
Dolostone†	Dolomite
Travertine	Calcite or aragonite
Gyprock	Gypsum
Anhydrock	Anhydrite
Rock salt	Halite
Chert†	Submicroscopic quartz
Chalk	Calcite or aragonite

*Some limestones are made up of fragments, either chemically or biochemically precipitated, that have been transported since they were formed. Such limestones, commonly termed *clastic* limestones, may be termed *calcirudite* (if predominantly gravel-sized fragments), *calcarenite* (if sand-sized), and *calcilutite* (if silt- or clay-sized).

†Many dolostones and cherts are diagenetic, having been formed by replacement of lime (either calcitic or aragonitic) mud.

sandstone plus an appropriate modifying term. Strictly speaking, sandstones may be made up of any mineral or group of minerals of sand size. Actually, however, most sandstones are composed predominately of quartz sand grains. Therefore, the term *sandstone*, unmodified, has come to mean "quartz sandstone" to nearly all geologists and others who deal with these rocks. As a consequence, two additional names for the other fairly common sedimentary rocks made up largely of sand-sized particles have gained rather wide acceptance: *Arkose* is a sandstone that contains 25 percent or more feldspar grains; it commonly resembles granite. *Graywacke* is an impure sandstone that contains more than 25 percent dark-colored mineral or rock grains; it is generally gray or greenish gray in overall color.

METAMORPHIC ROCKS

Metamorphic rocks are formed by the transformation of preexisting rocks. Igneous rocks, sedimentary rocks, and even metamorphic rocks may be metamorphosed or remetamorphosed, as the case may be. Metamorphic changes reflect responses of predominantly solid material to physical or chemical activities that are caused by high pressure, high temperature, or changes in chemical en-

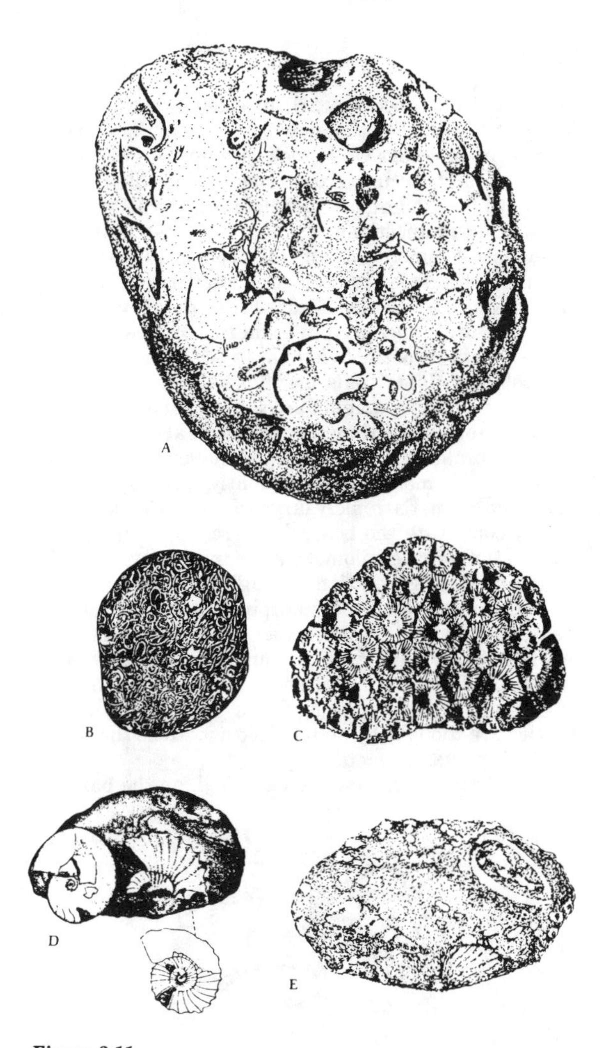

Figure 3.11
Stones' surfaces may expose fossils at any level and at any angle, in some cases making them essentially unrecognizable. In those shown, predominant fossils are as follows. *A*: brachiopods, *B*: pelecypods (clams, etc.), *C*: colonial rugose coral, the "Petoskey stone," *D*: cephalopods (nautilus), *E*: mixture of brachiopods, pelecypods, and chain corals.

vironment. The processes are directed toward equilibrating the rocks to the new conditions. As a rule, metamorphism occurs while rocks are buried fairly deep within the earth's crust. Exceptions to this rule include rocks that have undergone thermal metamorphism near shallow igneous injections or adjacent to lava flows.

The most common rocks metamorphosed in response to pressure processes are those brecciated and pulverized within fracture zones (see Figure 3.12); the metamorphism is generally termed **dynamic metamorphism**. Rocks transformed in response to heat or to activities promoted by percolating fluids (plus or minus diffusing ions) in zones around cooling magma are said to have undergone **thermal metamorphism** or **metasomatism**, respectively. Combinations of thermal and metasomatic effects around igneous masses are frequently termed **contact metamorphism**. Extremely large volumes of rock, which typically constitute the bedrock of areas measurable in thousands of square kilometers (or miles), are usually reported as having experienced **regional metamorphism**. Rocks in regional metamorphic tracts are thought to represent metamorphism under the influence of relatively high hydrostatic pressure (and, in some cases, also under the influence of differential pressures), elevated temperatures, and percolating fluids. Mountain-building activities are thought by many geologists to be the cause of such all pervasive metamorphism.

Metamorphic rocks may be classified, on the basis of

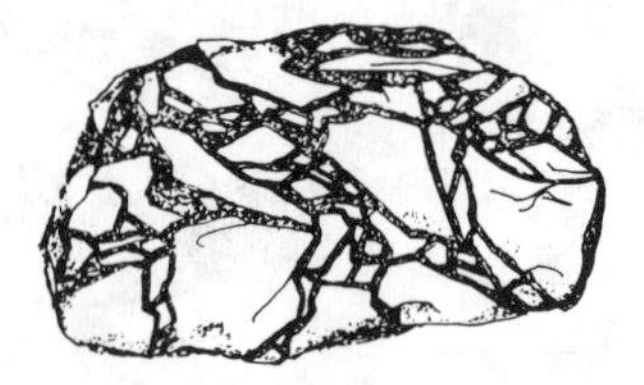

Figure 3.12
Some stones made up of fracture breccias can be distinguished from stones made up of sedimentary breccias because adjacent fragments in the fracture breccias may match each other like pieces of a jigsaw puzzle.

the arrangement of their constituent minerals, into two main categories: **Foliated metamorphic rocks** are those that have a readily recognizable preferred orientation of their tabular or platy mineral constituents (see Figure 3.13); **nonfoliated metamorphic rocks** are those that do not have such an arrangement of their constituents. The foliation may or may not be parallel to preexisting bedding or other layering.

The most common foliated metamorphic rocks are described briefly in Table 3.4. The pronunciations of their names are: gneiss—nīce, amphibolite—ăm·phĭ′·bō·līte, schist—shĭst, phyllite—fill′·īte, and slate—slāte. These names, except for slate, are often modified by a mineral or rock name. Examples are granitic gneiss, garnetiferous amphibolite, muscovite schist, and graphite-bearing phyllite. They also may be modified by descriptive terms such as *augen* (German for "eyes"), which is frequently used to describe gneisses such as the one shown in Figure 3.14. Names of slates often include color adjectives, for example, *red slate*.

The most common nonfoliated metamorphic rocks are quartzite and marble. Less common ones include hornfels

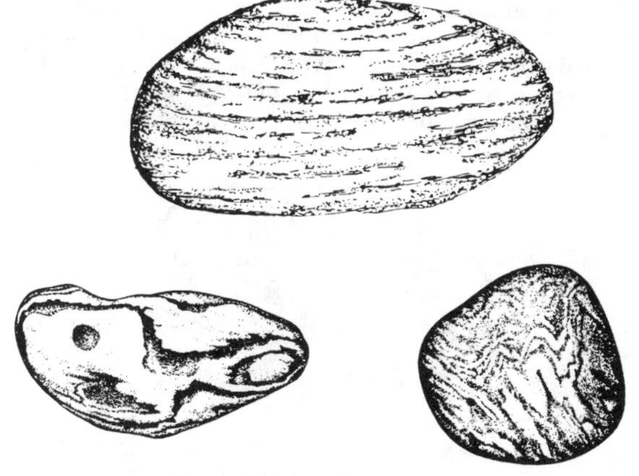

Figure 3.13
Gneisses are common as stones.

Table 3.4 Common Foliated Metamorphic Rocks

Name	*Features*
Gneiss	Imperfect foliation or banding; granular minerals—typically quartz or feldspars—predominate.
Amphibolite	Poorly to well foliated; green to nearly black amphiboles plus or minus off-white plagioclase predominate.
Schist	Well-developed, closely spaced foliation; platy minerals—commonly one or more of the micas or a chlorite—appear to predominate.
Phyllite	Intermediate between schist and slate; glossy luster; many are corrugated.
Slate	Homogeneous appearing; so fine grained that constituent minerals cannot be distinguished under a hand lens; can be readily split into thin slabs the planes of which need not be parallel to the original bedding.

and tactite. Some rocks generally considered to belong to the nonfoliated category are banded or layered and thus appear foliated upon only cursory examination. Such pseudofoliation, however, is not a product of metamorphism. Instead, it is inherited from its unmetamorphosed parent rock.

Unfortunately, the term *quartzite* has been used two ways: for silica-cemented sandstones and for metamorphosed sandstones. Therefore, the prudent geologist now uses the terms *metamorphic quartzite* or *sedimentary quartzite* as appropriate. Quartzites can be distinguished

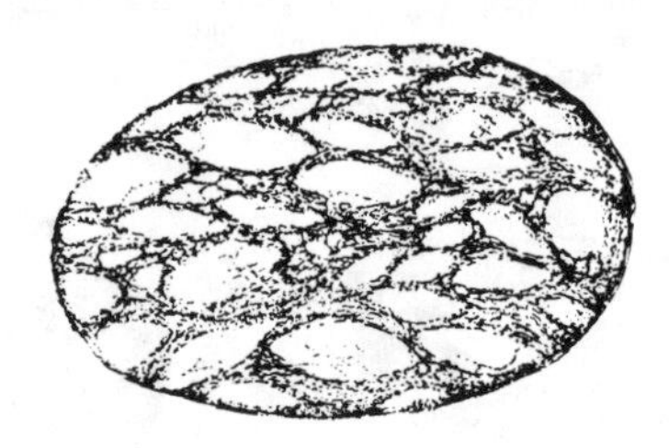

Figure 3.14
Augen gneisses contain feldspars, or less commonly other minerals, as relatively large eye-shaped grains.

from sandstones by looking at the fractures—fractures cut indiscriminately across the grains and matrix of a quartzite but around the grains of a sandstone. Indeed, the fractures of some quartzites appear conchoidal, just like those of individual quartz grains. (A *conchoidal fracture* is a seashell-like concavity formed when brittle materials are broken.)

The term *marble* has also had dual usage: In the commercial world, it has been applied to any predominantly calcitic or dolomitic rock that will take a polish; within the geological community, it has been generally restricted to metamorphosed limestones and dolostones. Metamorphic marble is characterized by interlocking grains. This characteristic alone, however, cannot be used as a criterion of metamorphic origin because the mineral grains of some unmetamorphosed, recrystallized limestones and dolostones are also interlocking.

Probably the easiest as well as the safest way to decide whether a marble or a quartzite is metamorphic or not is to determine whether, in its bedrock area, the rock in question is associated with other metamorphic or other sedimentary rocks.

Hornfels is the name given to rocks that have been metamorphosed in response to high temperatures near intruding magma. Many have hornlike appearances. *Tactites*, which also occur near intrusives, include minerals that contain elements introduced, generally in the gaseous state, from a magma.

Other metamorphic rocks also occur commonly as stones. Four examples are *soapstone*, massive metamorphic rock composed mainly of soapy-feeling talc; *greenstone*, greenish gray or green, either massive or slightly foliated rock composed largely of green amphibole or chlorite; and *epidosite*, a metamorphic rock that consists largely of pistachio-green epidote.

Metamorphic rocks are often mapped on the basis of the presence of one or more *index*, or *key*, minerals. These minerals are thought to be good indicators of the intensity of the temperature-pressure conditions under which the rocks were metamorphosed. For example, kyanite is indicative of high pressure; sillimanite, of high temperature; and chlorite, of low temperature and moderate pressure. Field relationships, laboratory investigations, and theoretical considerations have established the guidelines and restrictions basic to current hypotheses relating to metamorphism. The hypotheses will certainly be modified as more is learned about these most interesting rocks.

OTHER ROCKS

This catchall designation is used for rocks that do not fit well into any of the three main categories. Included are rocks formed chiefly as the result of diagenesis, pyroclastic rocks, migmatites, nonigneous pegmatites, and solid hydrocarbons.

Diagenetic Rocks

Diagenesis is the name given to primarily chemical processes that take place in sediments after their deposition but before, and commonly contributing to, their conversation into solid rock. Recrystallization and replacement are exemplary. *Recrystallization* is the formation of new mineral grains where others once existed; for example, fine particles in a calcitic or aragonitic mud might be dissolved and redeposited, essentially in place, to form relatively large interlocking grains of recrystallized calcite. The result in this example would be a nonmetamorphic recrystallized limestone, which is one of the rather common diagenetic rocks. *Replacement* is generally characterized as involving essentially simultaneous removal of one mineral substance and deposition of another in its place; for example, the replacement of calcitic or aragonitic sediment by dolomite or chert (cryptocrystalline quartz). It is now known, for instance, that most dolostones probably were not precipitated directly from seawater but were instead formed as a result of diagenetic activities whereby the mineral dolomite replaced a preexisting calcitic or aragonitic sediment. At least some, perhaps most, chert has also been formed as the result of replacement of calcium carbonate sediment.

Several other features commonly exhibited by stones may also have been formed as the result of diagenetic processes. Probably the most frequently recognizable ones are concretions, stylolites, liesegang banding, dendrites, and druses. *Stylolites* are irregular seams of in-

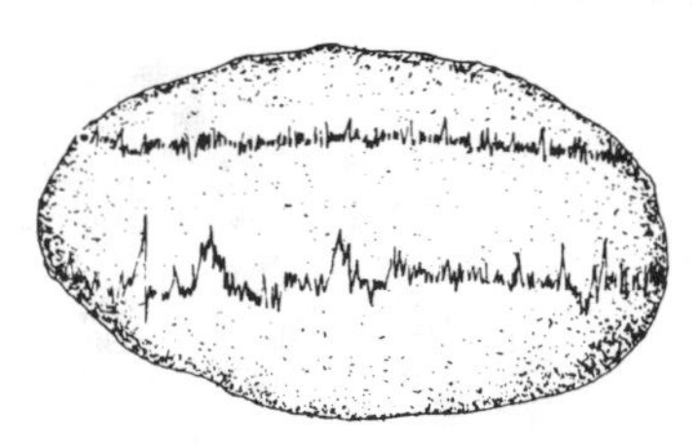

Figure 3.15
Stylolites.

soluble material, typically in limestones or dolostones, that are expressed as zigzag lines on planar surfaces (see Figure 3.15). *Liesegang banding* (see Figure 3.16) is color banding that is characterized by a nested arrangement—the layers are arranged concentrically like the layers of an onion. Some people have suggested that the banding is formed as the result of rhythmic precipitation within fluid-saturated rock. At least some liesegang banding appears to have formed as a result of chemical weathering; it parallels the surfaces of loose fragments. *Dendrites* is the name given to branching, treelike mineral growths that occur here and there on some rock surfaces (see Figure 3.17). These arborescent growths, typically black, consist of one or more of the manganese oxide minerals. *Druses*, which may be described simply as cavities lined with tiny, protruding crystals, may also be of diagenetic or later origin. Rocks and stones that contain fairly closely spaced druses are generally referred to as *vuggy*.

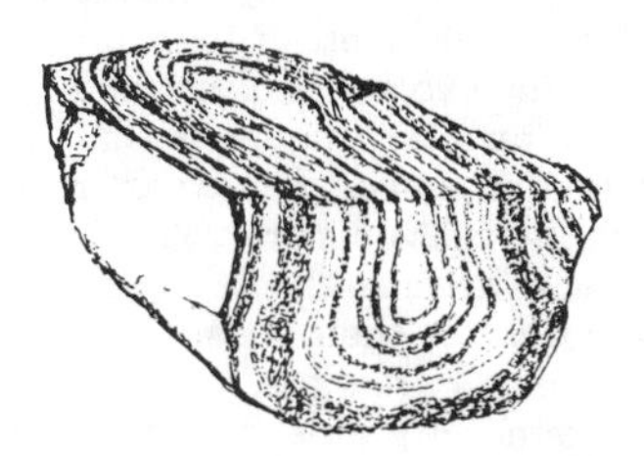

Figure 3.16
Liesegang banding.

Figure 3.17
Dendrites are most often seen on relatively flat surfaces.

Pyroclastic Rocks

Pyroclastic rocks consist of consolidated igneous fragments that were extruded explosively from volcanoes and then deposited, like sediment, by settling either on land or in water. Some geologists prefer to classify these rocks as igneous; others, as sedimentary. Perhaps the best suggestion is one generally attributed to C.K. Wentworth, a geologist who spent much of his professional career observing volcanic activity in Hawaii: "They are igneous on the way up and sedimentary on the way down."

In any case, pyroclastic rocks are usually classified on the basis of the size of their constituent fragments, just as detrital sedimentary rocks are. *Tephra* is the overall term for the deposits of loose pyroclastic fragments; *tuff* is the overall term for the consolidated rocks. Pyroclastic rocks composed of fragments less than 2 mm in diameter are called *ash tuffs*; rocks composed chiefly of pyroclastic fragments ranging between 2 mm and 64 mm in diameter are called *lapilli tuffs; pyroclastic breccias* or *block tuffs* are composed predominantly of angular fragments that are larger than 64 mm in diameter; and *agglomerates* or *bomb tuffs* are made up of 50 percent or more volcanic bombs. In most cases, these rocks are also classified and named on the basis of the nature of their constituent fragments: If most of the fragments are glass, the rock is termed *vitric*; if most are mineral grains, *crystal*; if rock fragments, *lithic*; and, if a general mixture of two or all three kinds of materials, *mixed*. Therefore, as an example, a pyroclastic rock made up largely of ash-sized fragments of glass would be called a *vitric ash tuff*.

Some special pyroclastic materials—volcanic bombs, Pele's hair, and Pele's tears—have been described in the preceding chapter. Pumice, another glassy product of explosive volcanic activity, is also rather frequently found as beach stones. Though typically light gray to off-white in color, pumice may also be yellowish, brownish, or slightly reddish. It consists of subparallel, entangled fibers of frothlike glass, in some cases surrounding sporadic crystals. Most pumice is so porous and light in weight that it floats on water. Consequently, it is likely to be found just about anywhere, even on beaches that are remote from volcanic activity.

Migmatites

Migmatites are intimate mixtures of metamorphic and igneous or igneous-appearing rocks (see Figure 3.18). In most migmatites, the obviously metamorphic component is a dark-colored amphibolite or biotite gneiss, whereas the igneous component is light colored and of granitic to granodioritic composition. Most migmatites appear to have been formed at relatively high temperatures when the light-colored material was relatively mobile, perhaps magmatic, while the darker-colored component was much less mobile, in most cases essentially solid rock.

Pegmatites

Pegmatites are, for the most part, composed predominantly of quartz and one or more of the feldspars, in most cases microcline or albite or an intergrowth of the two, commonly called *perthite*. As already mentioned, some pegmatites appear to be of igneous origin whereas others do not. Those that are not igneous have apparently been

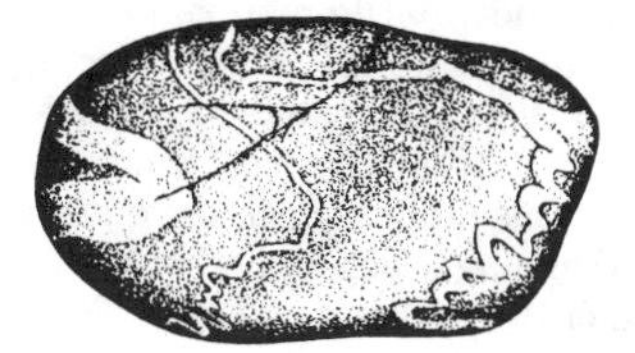

Figure 3.18
A migmatite with ptygma (tortuous, foldlike veins or dikes) of granitic material within a dark-colored amphibolite.

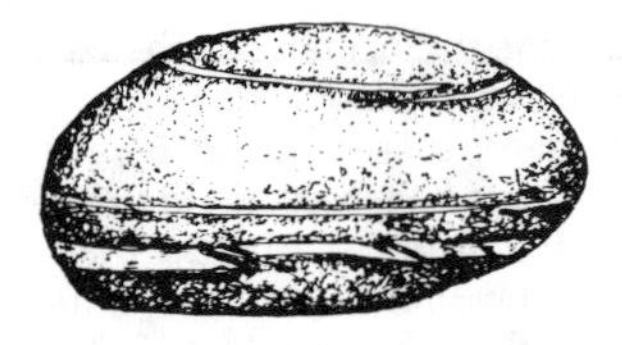

Figure 3.19
Quartz veins, fractured since formation, in a metamorphosed graywacke.

deposited by gases or hot water solutions and thus do not fit well into any of the three major rock classes. Most mineral veins (see Figure 3.19) also appear to have been formed as the result of deposition from hot water solutions. Many stones that consist of only quartz or pink feldspar have come from pegmatites or mineral veins.

Solid Natural Hydrocarbons

Solid natural hydrocarbons are of organic origin and, as already mentioned, are not considered to be rocks by some geologists. Nearly all of these hydrocarbons are so intimately associated with rocks, however, that it seems only proper to treat them here. *Peat*, which resembles compressed tobacco, is composed of only slightly modified plant materials. It represents the first stage in the change of vegetable matter into coal. *Lignite*, the next member of the gradational series between peat and anthracite, is a dull brown or brownish-black earthy coal. *Bituminous coal*, a brittle brownish or gray-black coking coal that typically is banded, is the soft coal of the market place. *Anthracite*, a vitreous black coal with conchoidal fracture, is the hard coal that is generally thought to have been formed by subjecting lower rank coals to the great pressures associated with major rock-folding processes. Some anthracite exhibits peacock-colored coatings on relatively fresh fractures. In the series from peat to anthracite there is a gradual loss of moisture and volatile material and a consequential gradual gain in the carbon content (commonly referred to as *fixed carbon*). The ranks of coal are based on these factors. Only bituminous coal and anthracite are likely to be found as stones. Both have been found in glacial deposits and among beach stones. It seems likely that many, if not all, of the coal stones found along beaches represent mined coal that has been dropped by ships.

Amber is another solid hydrocarbon that occurs as stones here and there, especially on beaches along the Baltic and North seas. Used as a gem and as the raw material for decorative items since at least 8000 B.C., it is fossil resin—that is, it represents ancient, partially devolatilized fir tree gum. Amber is reddish brown to pale yellow or rarely blue; it ranges from transparent to turbid; it has a resinous luster, conchoidal fracture, a hardness of 2–2½, and a specific gravity (~ 1.05) such that it sinks in fresh water but floats in salt water; it dissolves in organic solvents such as chloroform; it melts in a flame—in most cases yielding an oily substance and giving off dense white fumes that irritate the nostril; and, when rubbed with wool or silk cloth, it becomes charged with negative electricity so it will then pick up lightweight particles of paper, dust, and so forth. Some amber contains inclusions of such things as entrapped bubbles, insects, or pollen.

STRUCTURAL FEATURES

Another aspect of some rocks' histories and appearances falls into the field of **structural geology**, the branch of geology that deals with both original and imposed rock structures. *Original* structures are those formed while a rock is being formed; for example, during sedimentation or during consolidation of a magma. The geologist frequently uses such features to tell whether a rock unit that has been structurally disrupted is still right-side-up or has been overturned. *Imposed structures* are those formed when a rock unit is deformed; for example, during earth movements that accompany mountain-building activities. Both kinds of structures can be seen in stones (see Figures 3.20, 3.21, and 3.22).

Original structures that are relatively common in stones include mudcracks, ripple marks, crossbedding, and graded bedding of sedimentary origin and "pillows" of igneous origin.

Mudcracks are irregular fractures, commonly with a polygonal pattern, formed by shrinkage of silt or mud dur-

Figure 3.20
Crossbedding in a sandstone cobble.

ing drying. Those that exhibit curls are especially distinct in stones.

Crossbedding is layering that is at an angle to the overall stratification of the unit within which it occurs (see Figure 3.20).

Graded bedding is layering in which the grain size in individual beds grades from coarse to fine, bottom to top.

Pillow structures are so named because in cross section they resemble stacked pillows. Typically the pillows have tops that are convex and bottoms that accommodate the configurations of the surface upon which they were deposited. Originally, most pillows have glassy exteriors and a grain size that increases toward their centers. Most pillow lavas have been formed when lava has been extruded under water; some have been formed when lava has flowed over land, into water.

Imposed structures include folds (see Figure 3.21) and fractures (see Figures 3.19 and 3.22). The nomenclature of folds is given in most introductory geology textbooks;

Figure 3.21
Folds are found in some stones, especially those derived from metamorphic rock terrains.

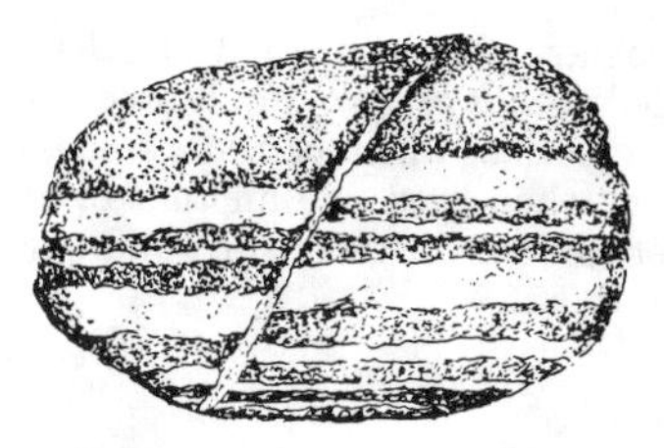

Figure 3.22
A fault. As oriented here, the left side has moved down with respect to the right side.

it is not given here because in stones the folds have been detached from their surroundings and thus, in most cases, cannot be named unambiguously. There are two kinds of fractures: *Joints* are breaks that exhibit only separation (movement perpendicular to the break); *faults* are fractures along which the rocks on one side have moved laterally with respect to the rocks on the other side (see Figure 3.22). Many fractures in detached pieces are also difficult, if not impossible, to name unambiguously. Therefore, just as was suggested for folds, an introductory geology textbook should be consulted to learn their nomenclature.

In some cases, jointing and faulting have left open spaces and, subsequently, fluids have deposited minerals in those spaces. Where the fluids were magmas, the resulting features became dikes or sills (see Figure 3.23). Where the fluids were gaseous or aqueous solutions, the fillings became veins (see Figure 3.19). Stones with included veins are rather common at many localities. In addition, many of the stones that are made up of only one mineral have come from veins.

In some fault zones, the rocks are broken up into an-

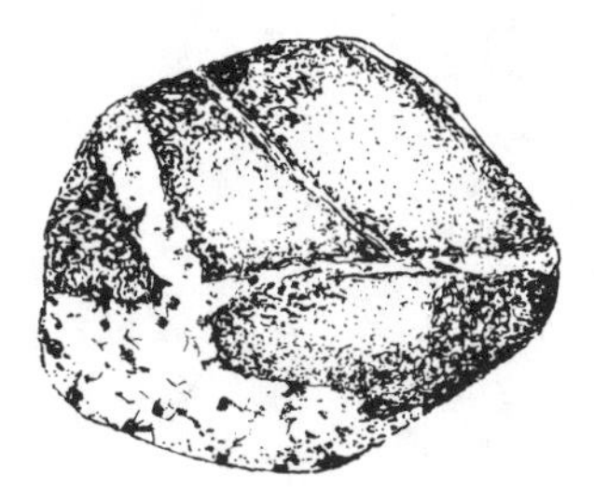

Figure 3.23
Dikes (the light-colored material) are filled with consolidated magma and thus are igneous rocks.

gular pieces. If the pieces are later cemented together, the resulting rock is called a *fault breccia*. Fault breccias and sedimentary breccias, both of which occur as stones, can frequently be distinguished from one another even after they have been detached from their surroundings (compare Figures 3.10 and 3.12).

PSEUDO-ROCKS

Rocklike materials produced for use or formed as a by-product of, for example, smelting operations are worthy of note because they occur along with natural rock materials at many localities where one is likely to collect stones, especially along beaches of major waterways such as the Great Lakes. Some pseudo-rocks have been water-worn to the point that their shapes closely resemble those of stones made of natural rock material. The manufactured materials that are most frequently found are brick, ceramic materials (including pottery and tile), cinders, coke, concrete, glass, and slag. None of these materials should be confused with stones that consist of natural rocks and minerals. (Brief descriptions of these materials are given in Dietrich and Skinner, 1979.)

4

Uses of Stones

Almost as long as any of us can remember, we have known that silver and gold, brass and steel, precious and semi-precious gems, and many other things are made from minerals and rocks. Books about economic geology and mineral resources deal with literally thousands of uses. In addition, nearly all of us have used or seen others use stones for such things as hammers, missiles, or weights to hold something down. Most of these kinds of uses, however, are not even mentioned in books about mineral resources.

This chapter briefly records such present-day uses of stones—that is, it tells of ways stones are currently used *just as they may be picked up*. For the most part, however, persisting primitive uses, such as flailing clothes on stones to help launder them, are not included. Neither are uses involving either large-scale recovery operations or fashioning processes; for example: riprap, which is quarried; flagstones, which require splitting; or poultry grit, the production of which nearly always requires crushing. Nor is the widespread use of stones as collectables included; that use has been noted throughout the preceding chapters and is also considered in Chapter 5.

To facilitate possible reference, the uses are presented under six broad headings: weights, sports, health, rituals, tools and construction, and miscellaneous uses.

WEIGHTS

Whenever I think of using a stone as a weight, I recall a line from one of Tom Lehrer's songs: "She weighted her brother down with stones and sent him off to Davy Jones."* R. A. Salit of Altadena, California, has reminded me that in a similar vein a gunny sack with stones in it, along with a litter of kittens or puppies, dropped in a stream or lake is still widely used—morbid though it is—for "farm pet population control." In any case, it seems likely that stones have been used as weights ever since human beings made their appearance on earth. Application of the term *stone* to a widely recognized weight unit, now generally accepted as equal to 14 pounds avoirdupois, underlines such use.

Even today, when most needs for weights are fulfilled by manufactured items, stones are still being used. This is true because here and there, now and then, stones fill on-the-spot temporary needs, cut down on costs, or best satisfy some aesthetic desire. Each of these uses depends on a stone's being heavy enough to hold something down, to impede movement, or to impel some gravity-driven mechanism.

Relatively small cobbles find frequent use as paperweights. For example, at the public display in Scotland's Inverary Castle, natural stones held down writs sent by British sovereigns to different dukes of Argyle. Today, stone paper weights with painted designs constitute a more than million-dollar-per-year business. And similar stones bearing either decoupage, or sandblasted "etchings" or hammered metal designs are becoming more and more popular.

Large cobbles or small boulders may be found as doorstops in humble cabins, average homes, and luxurious mansions. Some interior decorators have gone so far as to advertise that they carefully select such stones to fit with different decors.

The drummer with the modern jazz band Matrix uses a basalt porphyry cobble on top of a quilt inside his bass drum both to hold the quilt in proper position—an addition that gives the drum the desired resonance

---

*"The Irish Ballad," published in 1952 by Tom Lehrer.

qualities—and to help keep the drum in place while it is being thumped.

One manufacturer of ancient clock reproductions suggests that stones be used as the weights if authenticity is desired.

Many multistoried beehives throughout the world are weighted down by cobbles or small boulders (see Figure 4.1).

A Norwegian friend, Maureen Otter, has written, that stones hanging on ropes are used to keep the thatch from blowing off cottage roofs along, for example, the windy Donegal Coast of Ireland.

Siberian Eskimos construct their mazinka houses of walrus skins that are stretched over whalebone frameworks and are held in place by the weight of boulders hung from the skins by either leather thongs or sinew.

In Scandinavia, in particular, large cobbles and small boulders are sometimes put along the edges of sod roofs to secure the birch bark that is placed over the rafters,

Figure 4.1
Beehives are weighted down with stones in many countries throughout the world.

beneath the sod. Also in Scandinavia, and sporadically elsewhere, small boulders are used widely as weights atop chimney-cover slabs.

Quartz boulders are carefully selected for use as the weights in kraut barrels, probably because quartz does not react with lactic acid. Both vein quartz and quartzite boulders are used.

Washed clothes on, for example, the Canary Islands are often dried by spreading them out on warm stones on beaches. The clothes are kept in place by putting stones on top of them.

At least some of the soybean paste produced in Japan has as one of the steps in its preparation an 18-month fermentation in cedar barrels. During that time, the beans are compressed by stones—200 or more boulders and large cobbles per barrel!—that are carefully stacked in roughly conical piles on top of them.

Itinerant fishermen sometimes use boulders as anchors and pebbles as sinkers and some commercial lobster fishermen weight their pots (traps) with locally available, generally rather flat, stones.

In Morocco, for example, donkeys are often tethered by having a rope or thong tied around one of their front legs and the other end tied around a boulder of available rubble. Goats in the region are similarly tethered to control their grazing areas within yards.

A few pearl divers continue to use stones to help them descend.

Zircon mining in Sri Lanka (formerly Ceylon), according to the notes for the Colored Stones course of the Gemological Institute of America, involves bringing both the gem-bearing clay and water to the surface in buckets attached to winches that utilize stone counterweights. Also, in a few scattered areas, some primitive operations still make use of stones as counterbalances on treadmills.

Although cobbles and boulders are no longer used as ballast for cargo ships, pebbles are still used as ballast in the toy boats played with by the Nubian children of the upper Nile River Valley. The weight is needed to increase the stability of these little boats because they are made of metal.

Large cobbles or small boulders are sometimes placed on the downhill sides of vehicle wheels to back up brake systems.

When roads are icy, some people carry gravel in the trunks of their cars to increase the weight and thus increase drive wheel traction. In some instances, the gravel serves doubly because handfuls may be strewn on the slippery ice to increase the traction of spinning wheels. And the following story about the similar use of stones to free a vehicle stuck in loose sand has been related by C.B. Hunt (personal communication, 1979): "A geologist with one of the talc mines in Death Valley got stuck in sand going home for the weekend, so he loaded the back of his pickup with stones. He got free and when he arrived home in Los Angeles, decided to try to dump his unwanted rocks at a nursery. He asked if they could use some Death Valley rocks. 'Sure,' came the reply, 'I'll give you $25 for the load.' He soon broke that market."

Wire mesh rock catchers along unstable road and railroad cuts are frequently weighted down by boulders.

Stones used as counterbalances on swinging gates are fairly widespread (see Figure 4.2). In the first edition of *Stones: Their Collection, Identification, and Uses*, the sketch of the gate with the boulder counterweight had the gate hinged at the wrong end. When this was noticed, the sketch was modified for later printings to show a hinge-end reversal and a somewhat different, but also

Figure 4.2
Stones are sometimes used as counterbalances on gates.

rather widely used, setup (see Figure 4.2). Nobel Laureate Linus Pauling subsequently wrote and directed my attention to the fact that even this setup serves little, if any, counterweight function *per se*. As he remarked, merely tying the rope to extend between the end of the gate and the loop atop the long vertical member would give essentially the same effect. Answers to inquiries of owners of such gates as to what they think the setup accomplishes indicate that the arrangement is of "hand-me-down" origin, and that none of them had ever really thought about its function or lack of such.

Small boulders are frequently attached by wire, chain, or plastic cord to the bottom strands of open mesh fencing where it has been stretched over fairly long spans. The purpose, of course, is to weight the fencing down so there will be no large gaps beneath it where a cow, a steer, or a sheep could get out. Examples of this use are most common in areas where washed-out banks or small stream valleys are present along fences around grazing areas.

Here and there—for example, in northern New York and in Michigan's Upper Peninsula—where the overburden is very thin, farmers hold snow fence and even permanent fence posts down by hanging boulders on them. On a farm near Grafton, Ontario, also where the overburden is thin, small boulders are used as weights in the crotches of short-rail tripods that have been placed at the junctures of a long-rail fence.

SPORTS

Stones have served several different purposes in games for amusement, games of skill, and the so-called sports that lead to the kill. For the most part, readily available stones have been picked up and used as missiles. Less commonly, some particular stone or kind of stone has been selected for, for example, weight-lifting contests, to afford a surface with certain qualities, or to be used as stuffing.

The Chinese-originated game GO—which is reputed to be the oldest board game in the world—utilizes fingernail sized, disc-shaped, black and white stones, preferably chosen to constitute "matched sets."

Stones that are relatively flat are sought by children

Figure 4.3
Flat stones are used in hopscotch.

who play hopscotch (see Figure 4.3). If the stones will also draw lines, they are considered even better.

Disc-shaped stones, preferably with well-rounded edges, are best for skipping on water (see Figure 4.4). In fact, in a 1977 newspaper item, the Mackinac Island [Michigan] Stone Skipping and Gerplunking Club issued a plea for such stones for use at its annual Independence Day stone-skipping contest. It was noted in the plea that because of previous use, there is now a dearth on the island of "the good, flat, ovoid, smooth skipping stone so cherished for the tournament."

Bocce and curling now utilize artificially shaped stones or stone substitutes, although natural stones were originally used. It also has been said that the Iroquois may have sometimes used the nearly spherical stones found in potholes as the balls in their lacrosse competitions. In any case, there are authentic records that indicate a stone was included with the stuffing inside the leather covers of some of the Indians' lacrosse balls—apparently because they could be hurled with greater speed and accuracy than the typical, lighter weight ones. Can you imagine getting hit in the head with a stone ball about

Figure 4.4
Well-rounded, disc-shaped stones are best for skipping on water.

the size of a tennis ball but weighing between one and one and a half kilograms (some two to three pounds)?

The "folkloric Olympiad" held in conjunction with the annual Tiurai celebration of Tahiti includes an event that involves lifting a boulder that weighs approximately 265 pounds (see Figure 4.5). As an added handicap, the boulder is greased!

Tests of strength that involve lifting and hurling of the Dinnie Stones and the famous Inver Stone, respectively, are held periodically at the Braemor Games in the Scottish highlands. And, nearly identical events, using stones of about the same weights, are held at some two dozen "Highland (or Scottish) Games" held in the United States and Canada.

The Swiss and the Swedes have famous contests based on the distance or accuracy of stone hurling. At the Unspunnenfest near Interlaken, Switzerland, yeomen heave an 85 kg (~ 190 lb) boulder as far as they can. The current record, by the way, is a little over three meters (nearly ten feet). At the Folk Sports Olympiad on Götland, both men and women participate in a game called *varpa*, in which they toss softball-sized cobbles at a peg at a dis-

Figure 4.5
Greased-stone lifting in Tahitian Tiurai celebration. (Sketch based on a photograph by C. Molenhouse in *Sports Illustrated*, February 10, 1986.)

tance of approximately 20 m (~ 65 ft). The winner is the one whose stone lands nearest the peg.

Have you ever wondered how the phrase "within a stone's throw" originated?

An Oregon friend notes that Mrs. Leslie Pridey, from whose land many fine thundereggs have been collected, used to tell the story that when she was young, cowboys threw baseball-sized thundereggs at cattle, apparently to make them move.

Gauchos' bolas are still used in parts of South America for hunting as well as to aid in the herding of cattle. The typical bola consists of three ropes, each with a stone-weighted end, fastened together by their unweighted ends. The bolas are twirled and then hurled while revolving so as to entangle the legs of the animals or birds being pursued.

The sling has had a long and illustrious history. For example, it is reported in the Book of Judges in the Old Testament that "Among all these were seven hundred picked men who were left-handed; everyone could sling a stone at a hair, and not miss" (Judges 20:16). In truth, however, the sling was made famous by David because of his encounter with Goliath. In the past, pebbles or small cobbles were used as the missiles. Today, however, when elaborate slingshots are available in many of the larger sport shops, manufactured projectiles are suggested as better. Nevertheless, in the field, many slingers—no matter what their sling—still use natural stones wherever they can readily pick them up off the ground.

Sadly, the use of stones in slingshots as well as for throwing at those identified as foes has been part of the violence in, for example, the Israeli-occupied West Bank. I have been assured that, contrariwise, the hurling of stones by the British suffragettes was never at people, only at buildings.

In the standard southern history *The War of the Confederates*, E. A. Pollard reported that in June of 1861, in Missouri, "hard pebbles and smooth stones were substituted for shot." A friend has mentioned that he has done the same thing while hunting in a remote area.

Deadfall traps continue to be used by some hunters. These traps involve the falling of a boulder on top of an animal that has fallen into a prepared hole or pit.

In some of the Pacific islands, such as Tahiti, some of the fish traps are made of stones and wire mesh.

In the central and southern Appalachians, pieces of slate are a requisite part of the home-made turkey callers. The slate is rubbed across the protruding lip of a specially made hardwood (generally cherry or maple) box.

Many bean bags are not filled with beans. Instead, they contain very small pebbles as their stuffing.

The last use of this category, which was sent by Nick Claudy of the American Geological Institute, is difficult to classify. Is it part of sport (gaming?) OR is it nutritionally directed and thus health related? Whichever, in parts of Greece, it is traditional, upon catching an octopus to dash it repeatedly—20 to 30 times—against some convenient, smooth stone. This activity presumably not only kills the octopus but also tenderizes the meat. Subsequently, the carcass is hung out to dry "like long johns on a line"; its appendages are cut into convenient portions; the pieces are cooked over an open fire or grill; and the resulting delicacy bits are served, usually on toothpicks, along with ouzo.

HEALTH

Since time immemorial, several different minerals and rocks have been ascribed various beneficial attributes for helping people retain or improve their state of health. Some of the uses have sound physiological bases; a few may be malevolent; others appear to be of little or no consequence except in the minds of their users.* One use of stones, however, has been found to be healthful for at least 5000 years: It is a needle-free adaptation of acupuncture whereby stones are rubbed on appropriate parts of one's body to relieve tension and/or pain.

Pumice stones, still advertised in all sorts of magazines, have long been used to help remove bunions and calluses. In the past, they were also used to redden ladies' cheeks, despite the fact that pumice is glass and powdered glass can be extremely irritating.

Smooth stones of certain shapes are collected and sold as tranquilizers for "thumb therapy" (see Figure 4.6). These stones are sometimes called "worry stones." One that a friend bought had an included writeup that reads:

> Hold it between your fingers and feel your tensions ebb. The comforting touch of its smooth surface seems to induce tranquillity and serenity. . . . People of ancient Egypt, the Orient and Mexico recognized the therapeutic value of a talisman stone. Royalty counted its Worry Stones as symbols of wealth, health and good luck. . . . Your Worry Stone has been especially designed to fit snugly in your hand. For peace of mind and serenity, carry it with you always, ready to use any time. . . .

Perhaps more noteworthy, Robert Dober of Maple Heights, Ohio, carries a variant he calls a "religious stone"; a practicing punster, Dober's stone is *hol(e)y*.

Similarly, smooth stones are carried by some people who tend to stutter. They use the stone as a distraction device, which frequently is all that is required to let them talk normally.

Because they rattle, hollow concretions that contain loose materials have long been used by mothers to entertain their children and also by witch doctors to punc-

---

*As the Hindu proverb indicates: "If faith is put in a stone, it takes care of anything; if not, it remains a mere stone." (Original is "Vishwas karó tó bhagwān, nahi tó pathar.")

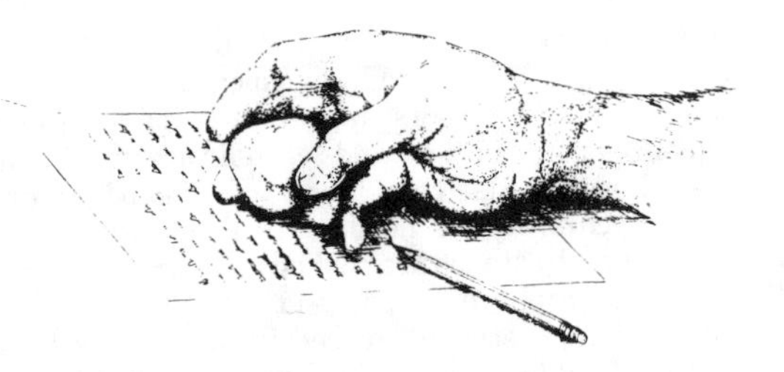

Figure 4.6
Smooth stones, when rubbed with one's thumb, may serve as tranquilizers.

tuate their rituals. The German name for these stones seems especially fitting—*Klapperstein*.

Although Demosthenes' placing stones in his mouth to stop his stuttering was probably only a distraction device, rather than of any physiological help, it is well known by most outdoorsmen that holding a pebble under one's tongue may cause salivation and thus help offset thirst. A number of hiking groups recommend the procedure.

Stones are used as a basic part of the Finnish sauna. Each must be selected so that it will not spall or desiccate upon heating up to approximately 120°C (~ 250°F). Stones that spall could cause serious injury to the sauna occupants. Stones that desiccate would soon disintegrate and subsequently serve no useful purpose.

Somewhat reminiscent of Zeus' punishment of Sisyphus, newly recruited miners at some mines in South Africa are required to shovel stones from bin to bin and back again.* The purpose, of course, is quite different; it is to help the men condition themselves before they assume their underground duties. Many of these men are physically "soft" because they come from tribes in which the women take care of essentially all of the physical labor.

---

*Mythology has it that the Underworld punishment of Sisyphus, the King of Corinth, was that he push a large stone up to the top of a hill, but that each time, just before reaching the top, the stone would roll back down so he would have to start pushing it up again . . . *ad infinitum*.

Large, generally tabular stones are sometimes placed upright in pastures for use as rubbing or scratching stones. This use is especially common in regions where many sheep graze.

Small stones, as well as coarse sand, serve as the grinders within the gizzards of many birds. Some of them become extremely smooth, and those from the large flightless birds of New Zealand and Australia were, at one time, eagerly sought by the aborigines as charms and ornaments. Also, for many centuries, human gallstones, kidney stones, and other calculi have been known, recognized as malevolent, and yet collected and carried as "charms."

Apparently in accordance with a tradition that dates back to at least the Dark Ages, some falconers still attempt to use stones as purgatives for their falcons.

Stones used in the drainage fields of septic tanks, in cesspools, and in drain barrels affect people's health in that they affect their environment. In many places, stones of specified sizes are commercially available, either bagged or in the bulk. At least part of this preparation and merchandising has been prompted by the implementation of rigorous zoning regulations relating to septic effluent disposal. The specifications, by the way, appear to represent spin-offs from the long-time use of gravel and stones for filtration in municipal and industrial water and sewage treatment plants, where the sizes and size distributions of stones that best serve the purpose have been established empirically.

Although stones that are used to line wells and reservoirs fill a completely different role, they also may affect people's health.

For untold centuries, stones of certain shapes, colors, or compositions have been used in several different ways because of beliefs that they would ward off disease. Some stones continue to be used as phallic symbols, as supposed aphrodisiacs, and even as presumed reducers of fertility. And, in one part of India, fossil plant impressions on stones are still scraped off and administered to people who suffer rheumatism.

Some Eskimos cook stones in water to make "stone soup." It seems likely that at least some of the stones they use are lichen-coated and that any nutrients and/or taste the soup may have must come from the lichen. Alternatively, as Arizona Bob Kubista has implied, in reference to "Indian Soup Stones," perhaps the mineral content of some stones flavor the soup. If this is true, one can not help but wonder: Should a stone that would so react with water be so-used, and how exactly might such a stone affect the brew?

A Public Broadcasting System television show widely aired in 1983 showed Hupa Indians of northern California using heated stones in basket "pots" to cook (i.e., to heat) their acorn soup.

The Polynesians on volcanic islands in the Pacific also cook with stones. Large basalt cobbles, cooking stones, are heated in underground ovens, called *imus*, along with fruits, vegetables, and meat. And, when whole pigs are cooked in imus, cooking stones are usually placed within the pig's abdominal and thoracic cavities as well as around the carcasses.

Small boulders are used in many ovens in which bread is baked, for example, in rural districts in southern Spain. The role of the stones is apparently one of holding heat within the appropriate temperature range.

Health-related uses for stones also involve animals: One example has been especially well documented by the British wildlife biologist Ian Redmond, who has reported elephant geophagy—that is, their eating of rocks and stones. The apparent impetus is the elephants' requirement for salt in their diets. Redmond actually observed forest elephants using their tusks as chisels to mine mirabilite (i.e., glauber's salt)-rich rocks within Kitum Cave in Elgon National Park in Kenya. In addition, the elephants' droppings commonly include stones up to a couple inches in diameter. These latter stones probably represent fairly insoluble rocks ingested along with the soluble salt-rich rock, and they may have been rounded within the elephants' digestive tracts.

Another example involves the use of stones as intrauterine devices to preclude pregnancy of camels while they are on long-term treks.

An apparently less valid use has also been jokingly (?) attributed to camel husbandry: The story has it that male camels may be made to add an extra gulp or two of water to replenish their stores—and thus extend the period during which they will not have to drink—if, when near the end of their pre-trek drinking, they are whapped just so with two cobbles.

RITUALS

From ancient times right down to the present, stones have served all sorts of roles in both religious and secular rituals. These rituals range from the proverbial ridiculous to sublime. Contrast, for example, the Blarney Stone ritual.

> There is a stone there
> That whoever kisses,
> Oh! he never misses
> To grow eloquent.
> Father Prout,
> in *The Groves of Blarney*

with Īd al-Aḍḥā (the Feast of Sacrifice), perhaps the most important of all sacred Muslim feasts. During this feast each pilgrim tosses forty-nine pebbles at three pillar "satans" that are located in Mina, near Mecca. Or consider the fact that Muslims are free to use stones, when water is not available, to wipe their hands in performing ablution before praying.

Some South Pacific natives still participate in age-old fire-walking rituals that involve walking on stones that have been heated over open fires. Today, however, at least some of the performances appear to be largely for the benefit of tourists.

Some Apache men still take *ritualistic* sweat baths during which their medicine men lead appropriate religious chants while sprinkling water on heated stones, usually four in number. These rituals usually take place in canvas-covered frames—Apache "saunas"—that are erected beside streams. The men also use their "saunas" for the purpose of purifying their bodies and their minds.

Both natural and carved stones often serve as objects of worship or to commemorate an event or the life of someone who has died. Several stones have been adored, some kissed, others saluted, and still others anointed with oil. It is noteworthy that many of the holy stones, particularly in western Asia, are meteorites. Their sacred quality was apparently perceived on the basis of their having fallen from heaven. Holy stones mentioned in the

Bible include Jacob's monument on Mount Gilead (Genesis 31:45–48) and the sacred stone of Ephesis (Acts 19:35). The sacred black stone in the Kaaba, the Islamic shrine in Mecca, may also be a meteorite.

Plymouth Rock is the best known commemorative stone in the United States of America. The Cumberland stone in the Battlefield of Culloden, overlooking Moray Firth just northeast of Inverness, Scotland, is another well-known commemorative. There also are a number of boulders in that battlefield serving as crude gravestones for clansmen who were killed there. In the new part of the main cemetery in Oslo, Norway, there is a fine example of the tasteful use of natural boulders as gravestones. The overall effect is that of an attractive rock garden. Several other cemeteries around the world also contain natural boulders as grave-markers; they are particularly common on graves of people who died in Nazi-occupied areas during World War II.

Although Stonehenge and other cromlecks, dolmens, menhirs, and trilithons belong to ancient history, cairns are still used rather widely as milestones and to mark boundaries, special locations such as a mountain top or the Arctic Circle (see Figure 4.7), caches, or depots of provisions. And, as is well known, many of these piles of stones are built up with considerable secular ritual.

Figure 4.7
Cairns are frequently employed as landmarks.

Susan Robinson, a well-known Canadian artist, has written that Eskimos in the flat-land areas of the Canadian Arctic often pile up rocks to resemble people. The piles are called *inuksuks*, a term usually translated as meaning landmarks. These inuksuks are used either singly or in clusters for such things as herding caribou (they frighten the animals!) and for directional landmarks, especially in barren plains.

An especially interesting cairn was erected on the crest of Mount Fuji to observe the once-in-a-century day 8–8–'88: Stones used for the cairn were brought in from 216 mountains the names of which include *fuji*. Mount Fuji was chosen as the site because its graceful slopes are said to mirror the Japanese character that represents the number 8—that is, two vertical lines that widen toward the bottom. The last stone was put in place at 8:08 A.M. by Akinori Yasuda, a boy who became 8 years old on 8–8–'88.

As one might suspect, in some places there is a recognition problem relating to groups of stones—for example, circles of stones. It is extremely difficult to determine if some individual boulders or clusters of boulders or cobbles have gained their present positions naturally or as the result of human efforts.

A travel brochure furnished travelers to the Andes reports that pebbles are used in traditional courtship practices of the Otavala Indians of northern Ecuador. Apparently the youth throws a pebble at the girl he wishes to wed and she, if she wishes to accept his offer, tosses the pebble back at him. One wonders if this is a harbinger of their future.

One of our sons, a minister, informs me that loose twinned crystals of staurolite (by definition, pebbles) with good cross shapes are still used as baptismal amulets in some areas in Switzerland.

As recently as 1977, four Saudi Arabian men were publicly stoned to death for kidnapping and raping a woman. The press release noted, "The four were punished in accordance with the Book of Allah to pay for their heinous crime. . . ." Unlike some of the biblical stonings, when large boulders were dropped on one or more persons placed in a pit, the recent stonings involved the throwing of small cobbles and pebbles at the victims. The stonings took place in the public square, apparently to be sure they would serve as an example for other would-be offenders. Also in Saudi Arabia, current law holds that adulterous women are to be stoned; beheading by a scimitar's blade is considered far too noble a death for such a misdeed!

At least two burial customs involve the use of stones: (1) The Kafirs of northwestern Pakistan use stones to "bury" their dead. Despite the fact that they are surrounded by Islam, these people have retained a religion that involves praying to wooden idols and worship of their ancestors and nature. As a consequence, they put their dead in carved wooden coffins, place the coffins on top of the ground, and cover them with stones. (2) Some Tibetans have "sky burials": Corpses are chopped into small pieces; the bones are smashed by using heavy stones; the fragments are fed to lammergeiers (bearded vultures). This custom apparently arose of necessity. Within the region where it is practiced, true burial is essentially impossible because of the terrain and cremation is limited because wood is too scarce to be used for such a purpose.

Not really a ritual, and very possibly a ploy made up by souvenir shop owners, the following may become a ritual and, in any case, may be a legend in the making: It is frequently reported that many visitors who have taken stones from the Hawaii Volcanoes National Park have found that they subsequently have had bad luck "just as an ancient Hawaiian legend predicts." According to the "legend," Pele, the goddess of volcanoes creates all sorts of problems for anyone who has the audacity to remove a stone from her realm. It has also been reported that, after having such bad luck, many visitors return the stones and that several of the containing packages include pleas to Pele for forgiveness.

TOOLS AND CONSTRUCTION

The use of stones as hammers was mentioned in the introductory remarks of this chapter; this use dates back to very early times. In fact, it is the most ancient use of stones by humans yet to be recognized. Among the 1.6 to 2.5-million-year-old artifacts found at the Koobi Fora site in northern Kenya, there were a number of well-rounded, apparently water-eroded, stones whose ends were battered. These stones, called *hammerstones* by their discoverers, are believed to have been used by early humans to flake the rocks that were fashioned into their crude cutting tools.

The reciprocal use of stones as anvils is almost as well known as their use as hammers. A documented example involves the use of the 1400-pound (4 feet in diameter) *Ring* fragment of the *Tucson Meteorite* as an anvil. It was apparently partially buried so a massive flat portion of the fragment could be easily so-used (Figure 4.8). Tom Bredlow, a well known ironsmith artist, is quoted as saying: "Anyone who ever worked iron, . . . who went out into the desert and came across the Tucson Meteorite, would know in an instant that it was heaven sent to be an anvil!" And, indeed, it seems that it was used as such by several frontier armorers and blacksmiths, even before the Arizona area became a United States territory.

Figure 4.8
A natural anvil, the partly buried *Ring* fragment of the *Tucson Meteorite*. (Sketch based on a photograph in *The Tucson Meteorites*, R. R. Willey, Washington, D.C.: Smithsonian Institution Press, 1987.)

A more widespread use of stones as anvils is one that many of us have employed. Stones that are handy are often used to crack nuts on as well as to crack nuts with.

A logical extension of the use of stones as hammers and anvils is their utilization in milling operations. Such use has had a long history, both for grinding grain and for pulverizing ore to facilitate the recovery of its desired constituents. Although, in most cases, rocks have been fashioned especially for such use, in a few instances available loose stones have been and continue to be used.

Perhaps the most obvious, as well as the most economically important, past- and present-day use of stones is for construction, for example, for cobblestone roads, building foundations and complete buildings, stone walls, and a few picturesque stone bridges. The Old World has several walled cities and towns as well as stonewalls serving as the downslope keepers for many of the terraced fields in regions such as those on the steep slopes of the foothills of the Spanish Sierra Nevada. The core of the Great Wall of China, reputed to be the only man-made structure on earth that is conceivably visible from the moon, is largely stones. Certain areas within the glaciated areas of the northeastern and midwestern United States and of nearby Canada are noted for their handsome (as well as garish) cobblestone houses and churches (see Figure 4.9). Another fairly common use of cobbles and small boulders is in the construction of both inside and outside fireplaces. In addition, many brick and even frame houses have stones as trim accents. And, in some regions, tarred roofs are topped and decorated by strewn gravel, including pebbles and small cobbles. As a matter of fact, annual production of gravel in the United States alone currently amounts to about 2.5 billion dollars.

A less obvious, but equally widespread, use involves the placing of stones around the bases of vertical structural members, such as fence posts and cornerstones, that are set in loose ground. A prime example—in several areas of permafrost in Siberia, large cobbles and boulders are placed in cribs around utility poles in order to assure their standing upright. Stonehenge, one of the truly remarkable structures ever conceived, exemplifies this use. Construction engineers offer the explanation that stones, so used, serve to distribute the weight of the vertical

Figure 4.9
Cobblestone buildings are especially common in areas covered by glacial moraine.

structure over a much larger area than that of its base and its buried vertical sections, thereby spreading out and increasing the supportive soil pressure and consequently decreasing the tendencies for tilting or horizontal shift.

The use of pebbles in the manufacture of decorative slabs with exposed aggregate appears to be increasing every year. Architects frequently specify these slabs for walkways or even for vertical paneling, both for indoor and outdoor exposure. In addition, urns, vases, disposal bins, lawn and patio furniture, and even tombstones are made of such concrete. This use of stones, by the way, dates well back in time: In certain regions of southern Europe, ancient walkways and walls covered by pebbles and small cobbles in concretelike matrixes are rather common. Indeed, in some places, the stones were carefully selected as to color, shape, and/or size and fashioned into designs. For example, at the Alhambra in Granada, Spain, several walkways have "lanes" marked by parallel lines of elongate pebbles and cobbles, and a few walkways have areas exhibiting flowerlike designs made up of black stones surrounded by other stones or with pebbles arranged to give herring-bone patterned areas. One cannot help but wonder if imitation of these patterns will become an "innovative" architectural feature of the future.

In many places, boulders are used to deter traffic. Two examples are at entrances to relatively wide walkways where automobiles are denied entry and along hazardous roadways where large boulders are used as guardrails. A somewhat similar "use" involved the Afghan guerrillas' placing boulders in strategic positions in order to block roads and thus set up ambushes.

The use of pebbles and cobbles for ground cover is increasing. This use appears to have developed as a natural outgrowth of their use in rock gardens, as borders for drives and sidewalks, and beneath overhanging eaves. One of my acquaintances who very much dislikes lawn-mowing had his whole yard covered with washed stones from a nearby glacio-fluvial deposit.

Another example of stones put down as ground cover is their widespread use for protecting roadcuts in unconsolidated overburden, for example, those along Interstate 90 in eastern New York State.

Some of the stones used as landscape accents in New York City and Philadelphia have been stolen. According to a column in the January 28, 1985, issue of *Time* magazine, large numbers of cobblestones have been removed from playgrounds, street corners, and roadways in those cities and resold for use in private landscape projects. In fact, the number of stones taken had become so great by the time the article was published, that the New York Bureau of Highway Operations had already spent thousands of dollars for asphalt and labor to repave gaps created in one of its parkways. Consequently, this "highway robbery" led to the New York police to come up with a new charge—"grand theft, rock."

Similar use, and very likely pilferage as well, is also widespread in several other geological terrains where stones occur. For example, stones are common accents within decorative cactus beds in the southwestern United States and Baja California. Also, some of the Baja settlements have property boundaries outlined by small boulders, and some of the adobe dwellings there have cobblestone floors.

The uses in landscaping, in particular, have led to a couple curious industries: (1) the production of molded plastic "cobbles" and "boulders," and (2) the making of stones from left-over pieces of rock at, for example, quarries that produce dimension stones.

Plastic "not-stones" are being widely marketed for use in, for example, pools and fountains. An interesting variant—a cobble-appearing container that is fabricated for secreting house keys and hiding "as an extra key" among landscaping stones—is also being sold. [Do they work? One of my neighbors once turned over more than a few real stones before finding his fabricated key-containing stone!]

The making of stones at quarries is exemplified by the operation at one of the large-dimension rock quarries in the southeastern United States, a region where natural stones are relatively uncommon. Scrap pieces of granite are roughly tumbled to make them rounded, and these tumbled rocks ("stones") are marketed for use as decorative ground cover. Also, rocks with certain desired qualities—for example, pure white gypsum and marble—are similarly made into stones even within areas where natural stones are abundant. All of these, as well as screened and washed natural pebbles, are available bagged or in bulk at many landscaping and building supply depots.

The utilization of stones for docks, breakwaters, and shore protection structures (groins, revetments, and seawalls) is well known. Although most such structures are usually thought of in relation to their capability to stop erosion along seashores and shores of other large bodies of water, like the Great Lakes, similar structures have also been found to be effective along the banks of rivers that are subject to floods. In addition, the above-mentioned use of cobbles to cover roadcuts may, in many places, be intended to deter slopewash and gullying rather than to fulfill some aesthetic whim. For shore control, cobbles and small boulders are used both directly and in cribs and gabions. Not surprising to anyone who has watched waves break on stone-covered strands, recent investigations have shown that stones of certain shapes and sizes are particularly effective in absorbing storm waves.

Along this line, my wife showed me the following

passage—granted about an historical rather than a current use, but nonetheless noteworthy—in R. K. Massie's book *Peter the Great, His Life and World* (Ballantine Books, 1980):

> The flat marshy country around the Neva delta . . . was almost devoid of rock . . . For years, every cart, every carriage and every Russian vessel coming into the city [St. Petersburg, now Leningrad] was required to bring a quota of stones along with its normal cargo. A special office was set up at the town wharves and gates to receive these stones [see Figure 4.10], without which the vehicle was not allowed to enter the city. Sometimes when these rocks were greatly in demand, it required a senior official to decide the fate of every stone.

In fact, the world over, many dikes, irrigation ditches, and culverts, as well as the previously mentioned road-cuts and reservoirs, are faced by cobbles or boulders.

MISCELLANEOUS USES

The following uses of stones include applications in such diverse vocations and avocations as the so-called culinary arts, backpacking, the visual and performing arts, advertising, and fads. A few uses by animals other than humans are also noted.

Unlike the previously mentioned, but questioned, culinary use as a stock for stone soup, smooth stones are indeed sometimes used in a few elegant Japanese restaurants: Miso(fermented soybean paste)-flavored dumplings are served on the stones instead of on plates.

Stones are frequently used to mark trails. Different-sized stones are usually arranged to indicate directions and, in some cases, distances too (see Figure 4.11).

Flint stones for striking fire are still sought by eager scouts. Frequently the search is productive, whereas the attempts to use the prescribed method are not.

A ring of stones is often placed around a campfire to prevent its spreading. Stones are sometimes heated in campfires by outdoorsmen for two different purposes: Stone-boiling of water and soup is fairly common; the use

Figure 4.10
Early eighteenth-century tollgate at entrance to St. Petersburg.

of hot stones as bedwarmers is less well known. This latter use may very well date back to the use, during colonial times, of heated stones inside long-handled, metal bedwarmers.

Just a few months ago, a backpacker in our woods was using a stone for a pillow—just as Jacob is said to have done when he dreamed about the ladder to heaven (Genesis 28:11–12). The stone used by Jacob, by the way, is believed by some people to be the famous Stone of Scone that is now incorporated, as the Coronation Stone, into the throne that is used during British coronation ceremonies. A much more common use is one that nearly all hikers are familiar with—stopping and using a large stone for what might be termed sitting-room pauses (see Figure 4.12).

Washed pebbles, generally 2.5–5 cm (~ 1–2 inches) in diameter, are used as heat storage beds in many solar heating systems. Pebbles of granitic composition are reputed to work best.

Figure 4.11
Stones are often used as trailmarkers by hikers, woodsmen, and scouts, even to better advantage than they were by Hansel and Gretel of the well-known *Grimm's Fairy Tales*.

Figure 4.12
A stone for one's sitting room.

Some tailors and dressmakers still use small pieces of soapstone or chalk as cloth markers. Most of these markers, however, are not natural stones; rather, they are cut and shaped pieces of talc schist or soapstone or even sintered limestone dust.

Many public washrooms, even in four-star hotels, along the Costa del Sol of Spain have small bowls nearly full of thumb-size pebbles for soap dishes. Soap placed in such bowls dries rather quickly and only a few spots of it stick to the holder (pebbles); thus the soap lasts fairly long and is not messy when picked up.

Small pebbles grace the bottoms of fish aquaria the world over. A few aquaria have also been embellished by larger stones with peculiar shapes or attractive colors. Some of the pebbles marketed for such use have been dyed various colors.

Some stones ring when hit. Included in this category are stones of such diverse rocks as slate, fine-grained dolostone, travertine, alabaster, and the igneous rock called phonolite. The rock name *phonolite*, in fact, comes from the two Greek words meaning "sound rock," and prior to being so named, the same rock was called *Klinkstein* because of the resonant tones it produced when struck with a mallet. Stones with this property have been used as dinner bells by many people. It also warrants mentioning that the writer has a set of slate stones that were selected so as to comprise a one and a half octave, chromatic scale "xylophone."

The clapper in some cowbells and sleigh bells is a small pebble.

Pumice stones, already mentioned under the heading "Health," are used as abrasives in furniture finishing. They are especially good for sanding intricately lathed pieces like chair legs.

Small cobbles are sometimes used to polish unbaked pottery of the primitive genre. Smooth stones of fine-grained rocks such as basalt and calcilutite are preferred.

J. Thomas Dutro, Jr., of the U. S. Geological Survey has reminded me that many field geologists [including me!] frequently use stones to sharpen their pencils. It has long been recognized that proper rubbing strokes on smooth stones give the extremely sharp points needed for spotting localities on field maps and aerial photographs.

Similarly, some artists who use charcoal as their medium utilize sandstone, usually sandstone pebbles, for sharpening their crayons. They insist that such sharpening produces the most desirable surface.

Stones are widely used as mountings for *objets d'art* and statues. Probably the most famous is the Little Mermaid in the harbor at Copenhagen, Denmark (Figure 4.13).

On a different scale, the famous toy soldier collection in Malcolm Forbes' "American Castle" in Tangiers, Morocco, includes several cobbles as settings, and Maignon Daga, a Latvian-born sculptor of Minneapolis, Minnesota, mounts many of his small ceramic and stoneware sculptures on large pebbles or small cobbles.

Boulders—especially highly weathered limestones with irregular shapes, including cracks and/or cavities—have long served as foci in formal oriental gardens (Figure 4.14). This use has even been depicted on a 1973 Chinese postage stamp that shows the painting *Lady Watering Peonies, Stone Ornament*.

Several sculptors have used and/or depicted stones. Three examples are (1) Dame Barbara Hepworth, who collected stones, mounted them, and imitated them in both clay and marble; (2) Isami Noguchi, who depicted

Figure 4.13
The Little Mermaid has rested on a boulder in Copenhagen Harbor, Denmark, since 1913.

Figure 4.14
A highly weathered limestone stone as a focal point in a formal oriental garden.

boulders in, for example, bronze and used natural boulders as foci in some of his famous works—for example, those in the Chase Manhattan Plaza in New York City; and (3) Robert Long and Bill Nazan, who have arranged cobbles and boulders in groups to create such things as stone circles, spirals, straight and zigzag lines, and snakelike forms.

Stones have also inspired several painters, photo artists, men and women of literature, and composers of diverse kinds of music. There are, however, far too many examples of such oeuvre to list here. Several are mentioned in my article, "Rocks depicted in painting and sculpture," which is cited in the Suggested References.

Pebbles are sometimes incorporated in the outer layers of decorative candles. Cobbles have been used by Paul Welch, an artist in Traverse City, Michigan, as the central motifs of some of his multimedia sculptures. The English artist James Waud carefully selects beach stones of certain shapes and colors to use in his well-known mosaics.

In a similar fashion, one of the attendants of the Dali Lama of Tibet supervised and worked on the production of a magnificent "Wheel of Time," using different colored sand grains (small stones!). As described and shown in the August 15, 1988, issue of *Time*, this fabulous "wheel" consists of "a centuries-old intricate design of Sanskrit, flowers, animals, and 702 symbols for Buddhist deities."

A number of concretions have been recovered from the lake deposits of the Salton Sea Basin of southern California and used as decorative accents in rock gardens and yards in and around Los Angeles. Many of the concretions closely resemble some of the non-representational abstractions fashioned by certain twentieth-century sculptors.

An unnamed stone collector has recently been reported to have collected and mounted boulders on large tripods for display as garden sculpture. In 1977, the controversial sculptor Carl Andre placed thirty-six boulders, some weighing several tons, in a roughly triangular grouping in a park in Hartford, Connecticut, to fulfill an $87,000 commission by the National Endowment for the Arts and a local group. In an even lighter vein, in the art section of a 1972 *Time* magazine, an hourglass filled with pebbles rather than sand "for people who don't want to grow old" was noted as one of the *objets introuveables* then on display at the Louvre's Museum of Decorative Arts. This list could go on and on.

Plastic-coated as well as tumbled pebbles are used in costume jewelry. One kind of stone of particular interest in the jewelry trade is usually sold as "good-luck fairy stones" or "fairy stone crosses." The natural ones are cruciform (cross-shaped) crystals of the mineral staurolite. They are found here and there loose in residuum over slightly weathered staurolite schists.

Cobbles and pebbles are sometimes used as background pieces in store window displays and in illustrations used for advertisements. Although jewelry stores appear to be the most frequent users, several others including clothing and sporting goods stores also so utilize stones fairly frequently.

Several specialty houses are currently advertising fancy wooden match holders, each topped by a well-rounded pebble or small cobble to be used as the striking surface.

An especially fanciful item marketed by Miles Kimball of Oshkosh (Wisconsin) is an "Indian Weather Rock" that consists of a pebble, twigs, leatherlike thong, and wooden base plus "instructions for interpreting the weather . . . [for example,] If rock is wet—it's raining. . . . If rock is moving back and forth—it's windy. . . ."

During the fall of 1975, a California entrepreneur, Gary Dahl, introduced one of America's patented short-lived fads—pet rocks. Each "pet" was a beachstone that was sold nestled in excelsior inside a specially designed cardboard carrying case. A short time after the introduction of these stones, *Newsweek* magazine reported that several tons of the stones, each in its case and with a cleverly conceived "training manual," had already been sold. Sometime later—following the enterprises based on piggyback beauty contests, pedigree certifications, special beds, and even food for pet rocks—pet rock cemeteries were established in a number of communities, and it was reported in the news media that a town on Lake Superior where rampant coastal erosion was occurring had petitioned the federal government to have their shoreline area designated as the "National Burial Ground for Pet Rocks."

According to an item in the *Los Angeles Times* (September 30, 1981), a Whittier, California, lady was marketing black Mexican pebbles, each with a stuck-on star, as male (gold star) and female (red star) "Rock Stars." Apparently the multi-million dollar fortune made from "pet rocks" continues to attract imitators.

More to the point, at least moneywise, boulders have been used as legal tender for the last approximately 1500 years on the Island of Yap in the South Pacific. Some of the currency consists of boulders up to 12 feet across that weigh as much as a few tons. If nothing else, it would seem that the larger denominations of such currency would be burglar-proof!

Speaking of criminal activities, an anonymous note tied to a stone and left on my desk reminded me that, in the past, a threat (for example) was sometimes attached to a stone and then delivered by throwing the stone through the addressee's window. Such delivery was, it was written, foolproof at least so far as "getting one's message across."

Perhaps even more criminal, and certainly one of the most disturbing and obscene uses of stones of which I have become aware was the recently reported attempt to utilize cobble- and small boulder-size lapis lazuli rough for drug smuggling. The stones were hollowed out, and drugs were secreted in the cavities for transport from Pakistan through Netherlands customs.

A number of uses of stones by animals other than humans have been reported. One of the first ones mentioned in the literature does not appear to be supported by present-day observations: Pliny (Book X. XXXIII) reported that in high winds migrating birds may carry stones in their feet as ballast to make their flight steadier. On the other hand, the following uses and several others do appear to be well documented. Sea otters off the California coast use flat stones as anvils on which they break the shells from abalone and other mollusks in order to obtain the soft internal parts for eating. The otters do this by holding the stones on their chests while they float on their backs. Some observers have called these stones the otter's "lunch counter." Sadly, it has recently been determined that over time these activities appear frequently to cause the otter's chest to cave in, put pressure on its heart, and ultimately lead to its demise. Several birds (e.g., the European song thrush and jackdaws) break snails by using stones or hard ground as anvils. The Egyptian vulture has been frequently observed to use stones to break ostrich eggs in order to crack them and thus obtain their contents (see Figure 4.15). The male Adélie penguin gathers stones and presents them to his mate as part of the nest-building ritual. Caspian terns generally rim their nests with pebbles. Several other birds, such as killdeers, include stones among their nest-building materials. Perhaps the most peculiar stones that are used in this manner are the hairlike strands of volcanic

Figure 4.15
Egyptian vultures use stones to break ostrich eggs and thus obtain their contents.

glass, called Pele's hair. Each strand of this "hair" is, by definition, a stone (albeit a peculiar one). Some of the skylarks on Mona Loa make their nests almost wholly of of these stones.

Fishes and even insects are also known to use stones. Jawfishes line the entranceways to their subsurface homes with pebbles and seashell fragments, often giving their excavations the appearance of masonry-lined wells. Freshwater mollusks frequently have small pebbles trapped within their furrows, and some snails avoid desiccation by nestling under pebbles in their sand-coated channelways. Some mason bees choose stones or rock as the sites upon which they build their nests, and others use small pebbles to close off the passageways to their eggs, which they lay inside snail shells. The larvae of some species of caddisflies attach sand and small pebbles to their casings—in some instances as a building material, in other cases to weigh the casings down so they will not move with the water flowing by them.

Marian Whitney, professor emeritus of biology, Central Michigan University, was amazed that in the first edition I omitted the fact that "both biologists and paleontologists have *for decades* hypothesized dispersal of life—for example, of insect eggs and larvae, seeds, and even fish roe—within cavities of floating and subsequently remotely grounded stones made up of such rocks as scoria and pumice."

In a completely different vein, I have recently seen

three dogs that were trained to fetch and play with cobbles rather than with balls or sticks. One of them, a five-year-old German shepherd, had a most amazing trick: He would retrieve a stone, once identified with the person throwing it, from among hundreds of similar stones all under up to about 0.5 meter (~ 1½ feet) of water along a stony beach. Remarkable as these tricks are, one cannot help but wonder how long these dogs' teeth will last. Response to the mention of this use in the first edition suggests that the activity is fairly common. As an example, Jane Coven of Elyria, Ohio, wrote at length about her Jack Russell terrier and its favorite stone—a quartzite cobble that the dog once dug up in New Mexico. It seems that ever since then, the dog has continually played with that stone and only that stone. Among other things, the dog is able to find and retrieve its quartzite cobble even when it is tossed among similar quartzite cobbles beneath shallow water.

In the September 1979 Animal Behavior issue of *International Wildlife*, it is said that "Polar bears amuse themselves with stones, which they sometimes balance on their heads."

In what seems to me to be removed from mainstream thinking, Ann Chase, a self-described researcher in "noetic science," has written that her Kirlian photography shows "pulsating life energy in . . . [*inter alia*] stones."

More down to earth, our second son, a professional musician, reminds me that stones in streams may cause "interesting, quieting or disquieting, intriguing, hypnotic, . . . sounds." In this vein, he directs attention to Oscar Hammerstein, II's lyric "To laugh like a brook when it trips and falls over stones on its way. . . ." Even the famous Buson alluded to this phenomenon in one of his well-known haiku:

> Winter storm,
> the voice of the rushing waters
> is torn by the stones
>
> (translation by Tadao Okazaki)

A use by people and animals: For nearly 250 years, prospectors and geologists have been tracing boulder trains (see page 32) back to their sources with the hope that valuable ore, the mother lode, so to speak, would be found there. Some deposits have actually been found by this method. Recently, dogs—because of their acute sense of smell—have been trained to help with such tracing. And now there are several well-documented cases in which dogs have been able to sniff boulders buried beneath soil. One dog in Europe has been reported to have detected the presence of certain sulfide mineral-bearing boulders at depths of 12 meters (~ 40 feet)!

One of the more interesting "uses" (i.e., roles) of stones sent to me in response to my request—which, by the way, is reiterated in the next paragraph—relates, admittedly tenuously, to the age-old question Do stones grow?, to ritual, to goodwill, and also to stone collecting. A stone in the old city of Amersfoort, in the Netherlands, has so-to-speak begotten several stone progeny. According to a 1984 letter from Ben Brugmans, this boulder, "De Amersfoortse Kei," was placed on a pedestal in a parklike setting within Amersfoort because stones any larger than sand grains are extremely rare there. For many years thereafter, despite (or perhaps because of) its eminence, the stone became the object of many pranks—for example, wags from rival towns frequently moved it and put it in all sorts of unlikely places. More recently, however, the "kei" has apparently assumed a much different role: It has been joined in its park by several additional stones—stones transported or sent from towns and cities the world over. Furthermore, many of the kei's scions, often called "good-will ambassador stones," bear plaques proclaiming their origins, their donors, and/or other information.

Do you know of other uses of stones? If you do, I would greatly appreciate your writing to me about them.

Figure 5.1
Needed equipment: collecting bag, wrapping material (newspaper), notebook, handlens, maps, and first-aid kit (it's in the bag!).

5

Collecting Stones

The following bits of information and hints will be old hat to many collectors. They are provided to answer questions frequently posed by beginning collectors.

EQUIPMENT NEEDED

Strictly speaking, no equipment is needed to collect stones. The following things, however, will sometimes be extremely useful (see Figure 5.1).

A *collecting bag* (e.g., a knapsack) or a pail will let you carry more than you can get in your pockets; thus, if you carry one, you will not have to make trips back and forth to and from your car or other means of transportation as you collect. Get one that is relatively easy to carry and not so large that it is too heavy to carry when full.

Wrapping material, such as newspaper or paper sacks, should be carried for wrapping specimens so they are separated from one another. Despite the fact that most, if not all, of the stones you collect have undergone natural collisions and abrasion during their shaping, some of them may be broken or marred if they are not protected while you transport them.

A *notebook*—preferably either spiral-bound or well sewn—should be carried and used to record pertinent information while you collect. Such records will complement your memory. Record the date, the geographic location, the type of locality (e.g., glacial versus beach deposit), the kinds of stones collected, and any other information you consider pertinent. You also may want to add such things as field sketches of different sites.

An *underwater viewer* is useful when looking for submerged specimens—for example, along beaches. These viewers make it possible to keep water ripples from blurring your vision. They can be simply constructed by sealing a sheet of glass on a length of stovepipe of appropriate length. It is wise to cover the viewer's end of the stovepipe edges with something like polyurethane foam so no one will be cut by the device.

A *trenching shovel* is sometimes useful in areas of unconsolidated deposits such as stream gravels and glacial drift.

A *handlens* will be helpful, if not required, for identifying the constituents of many stones, should you want to name them by rock type. Many of the features necessary for identification of small mineral grains—for example, their cleavage—are difficult, if not impossible, to see without magnification (see Figure 1.3). A 10× lens is best for most purposes because it generally gives sufficient magnification but still gives a large enough field of view for one to see the relationships among the constituents. Handlenses can be purchased from many bookstores and hobby shops as well as from geologists' and prospectors' supply houses. You can carry the lens in your pocket or it can be hung from a chain (piece of rawhide or the like) and worn around your neck. The latter is preferred by many collectors because it makes the lens readily accessible and less likely to be lost.

A *hard hat* and *hard-toe boots*, preferably with nonslip soles, are well worth having if you collect in gravel pits or near steep banks, especially if others are also collecting there.

Topographic, geologic, and/or county road maps may be great time savers so far as locating sites. Federal and state geological surveys and, in many places, regional or county agriculture extension offices sell such maps or, at the least, can tell you where to procure them.

A *first-aid kit*, including a snake-bite package, and some kind of *insect repellent* should be kept available.

A *camera* is handy for recording such things as the appearance of a collecting site. A slide or a photograph may become extremely helpful so far as complementing your notes and reminding you of certain important aspects about a site.

Masking tape is carried by many collectors. It is useful for fastening loose ends of wrappers and has the advantage that one can write on it.

COURTESIES AND RULES

In collecting, as in most activities, common sense is a fundamental requirement. In addition, it is important to have a respect for nature and the rights of others.

Those things said, the number one rule is to be sure to obtain permission from the landowner before entering or collecting on private property. Most landowners will consent to collecting on their property if you ask them and tell them your purpose. Unfortunately, many collecting sites have been declared off-limits because of the thoughtlessness of a few individuals who trespassed on the property without first asking permission.

Access to, for example, gravel pits is likely to be restricted, in most cases because the owners are concerned about injury to collectors and, of course, their liability for such injuries. Sometimes, however, organized groups or even pairs are given permission to collect in such places even though individuals are not.

Collecting is prohibited in all national parks. Collecting may or may not be prohibited in the lands around national parks that are supervised by the Bureau of Land Management and National Grassland Regional offices. Collecting is prohibited or restricted in many state parks. Furthermore, even in these areas where individual collecting is permitted, commercial collecting is generally restricted. Consequently, it is imperative that you check federal and state laws that may be relevant before you collect in such areas. Check in particular the current interpretation of the Antiquities Law of 1906 (U.S. Code

Section 34, Statute L-255): That law seems to be undergoing reinterpretation continually and, in some places, may even be applied to collecting stones.

Although roadcuts provide excellent collecting sites, collecting from them, particularly those along interstate highways, is usually prohibited. Even if you plan to park in a designated rest area or elsewhere off the road, you should check with the state police or county sheriff before collecting within any highway right-of-way area.

Once permission is gained, always keep in mind that you are a guest *and* your behavior is often considered to reflect on collectors in general. The fact is, the way you behave may determine whether others will be allowed to collect at the site in the future.

To list a few courtesies: Do not trample or otherwise damage growing crops. Do not disturb livestock. Close any gate you open (and leave open gates open). Within pits and elsewhere, leave machinery, supplies, and equipment where they are—that is, do not even move them with the idea of putting them right back. Be neat—for example, never leave litter, and be sure to fill any holes you make so that others (or livestock) will not stumble in them and suffer injury.

Even though stone collecting is a relatively safe pursuit, certain precautions must be taken. A few seem especially noteworthy:

When collecting along beaches, wear clothes and shoes that you do not mind getting wet. The best intended resolutions not to venture off-shore are frequently broken when an attractive specimen catches your eye but is just beyond your reach. It is unwise merely to remove your shoes; instead, wear old shoes—for example, sneakers—and leave them on while you wade so you will not cut, bruise, or otherwise hurt your feet.

When collecting in areas that are subject to tides, always remain aware of the time, the tidal schedule, and the fact that tides may come in rather rapidly. Be sure to allow plenty of time for walking—*not* running—to high and dry land, especially when you are in coves surrounded by cliffs or high banks.

Stay a safe distance from all cliffs and banks. Many of them are unstable and subject to intermittent slides or slumping. Remember that you should keep your distance from both the top edges, where your weight might start a slide (with you included), and the foot areas, where slide or slump debris could fall on you and even bury you (see Figure 5.2).

Wear a hard hat whenever you are collecting where there is anything above you that might be dislodged and fall on you.

Learn to identify and avoid such plants as poison ivy and nettles, and such animals and critters as poisonous snakes, scorpions, ticks, and spiders.

Finally, try to be choosy from the very beginning of each collecting trip. Anyone who is not tends to load up

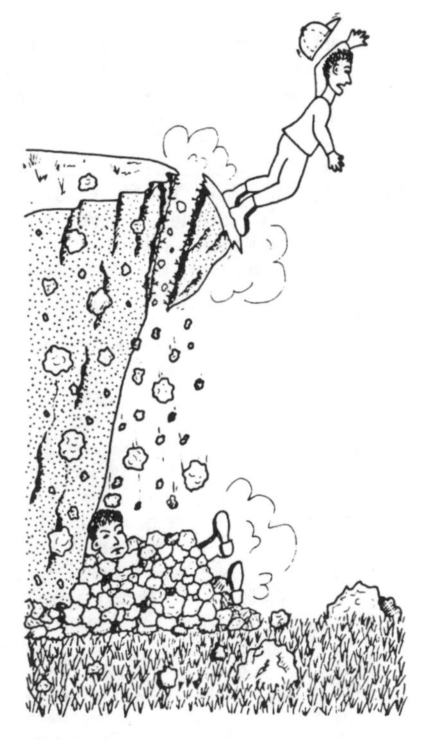

Figure 5.2
Keep your distance from banks and cliffs.

with many specimens that are later discarded. This is unfortunate—among other things, it precludes the possibility of someone else's having the opportunity to collect those specimens.

WHERE TO COLLECT

Stones can be found almost anywhere. This is true even though the local geology and the accessibility of collecting sites differs greatly from place to place (see Figure 5.3). Therefore, the road to success is knowing or finding out where to collect as well as what to collect and how to collect.

Although some good collecting localities are widely known and several others are easily spotted because of their names—for example, Stoney Point—probably the best way to find out about good places to collect is through contacts with other collectors. Many collectors are members of clubs, several of which are affiliated with the American Federation of Mineralogical Societies (920 S. W. 70th St., Oklahoma City, Oklahoma 73193); others may be found through inquiries directed to state, regional, county, and municipality tourist information bureaus. Most clubs hold monthly or otherwise scheduled periodic meetings. At the meetings, you will usually have the opportunity to meet people who will be willing to share their knowledge about such things as collecting sites and to help you in other ways. If by chance there are no such clubs in your area of interest, try contacting someone at a nearby museum, a high-school science department, or a college or university geology department.

In any case, before you go on a collecting trip, you should plan where you will go and know what you are likely to find. This is to say: A good collecting trip is almost always preceded by good pretrip planning.

A few general tips are well worth keeping in mind:

The best times to collect along beaches are soon after storms or just after the ice "moves out" in bodies of water that freeze. "New" stones are frequently uncovered by relatively strong storm waves or brought to the shore as the result of ice shoving, ice rafting, and other such activities.

The best times to collect along streams and on bars in

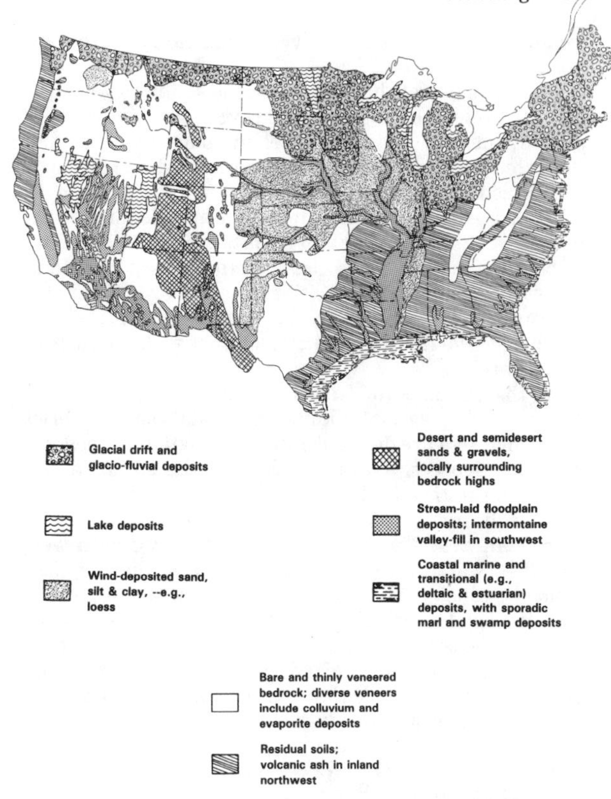

Figure 5.3
Generalized surficial materials map of the forty-eight coterminous United States of America. Similar maps are available for many other regions of the world.

streams are after periods of high water and relatively rapid flow, for example, after spring floods.

Stones in glacial deposits that are remote from the bedrock exposures from which they were derived tend to comprise a greater variety of rock types than stones in residuum or in deposits derived from nearby bedrock exposures.

If size is one of the considerations so far as what you collect, remember that many gravel pits have piles of stones that are sorted by size; in some of them, your preferred size may constitute the operator's discard ("happy to get rid of them") piles.

If you are particularly interested in collecting stones with diverse shapes, your best bet is to collect along streams or beaches that are fairly close to exposures of limestone, dolostone, marble, or formations that consist of intimate mixtures of two or more rocks, each of which has a resistance to weathering and erosion that is different from the others.

If the inclusion of stones of pseudo-rocks is a goal, areas where slag or building materials—such as brick and concrete—are dumped onto or just left along beaches are good sources. Two examples are the so-called slag beach on Big Bay de Noc on northern Lake Michigan and the beach area just west of Nerja (Spain) on the Mediterranean Sea. The former beach is covered with well-rounded, water worn pebbles made up of diverse slags from the nineteenth century smelter that was located at Fayette; the latter includes similarly eroded pebbles and cobbles consisting of all sorts of brick, tile, terazzo, mortar, and concrete intermixed with several natural rock types including marble that represents eroded remnants of leftover floor and molding slabs.

LABELING AND CATALOGING

Although stones without labels may be interesting or objects of curiosity, they almost always take on added interest if they are appropriately labeled. Indeed, the serious collector will label all specimens. Many collectors even number their specimens and keep a catalog that contains a permanent record of each specimen.

There are two kinds of labels: those attached to a specimen when it is collected in the field and those that are attached to the specimen permanently when it is put into a collection. Some collectors merely number their specimens in the field and record the number and information about the specimen in their notebooks. In my opinion, it is much better procedure to put more than just a number on a specimen in the field. If this procedure is followed, even if you were to lose your notebook, you would still have something more than just a collection of numbered stones and no specific information about them. In any case, whatever method you decide to use, field labeling is extremely important, especially when you collect at two or more localities on a given field excursion.

Each permanent label should include the kinds of information shown on the sample label in Figure 5.4.

In making a catalog, each specimen becomes an entry, generally with a number designation (± one or more letters), and the entries are usually made in an appropriate sequence both to increase and to facilitate their usefulness. Many collectors paint the designation on an inconspicuous part of the specimen so it can be correlated with the catalog even if it becomes separated from its label.

Each catalog entry should include at least the same data as the label. In many cases, it will also include additional details taken from field notes or laboratory studies. Some collectors include sketches or photographs of collecting sites and/or specimens in their catalogs. Some collectors have their collections cataloged in computer memory.

No. *42221*

Name: . . . . *granite (?)*

Loc.: *5 ft. above water level, South side of Chippewa Crk., -2.6 mi NW of Hammond, N. Y.*

Geol.: *Precambrian (Late?)—see Buddington, N. Y. State Mus. Bull. 296-map.*

Collector: *F. S. Dietrich*

Date Collected: *May 3*, ...

Figure 5.4
A typical label.

CLEANING AND POLISHING

Many stones are cleaned and/or otherwise modified by collectors, especially if their desire is to display them. Soap and water combined with a good brushing will often suffice. For some specimens, however, coatings of such things as lichen, algae, calcite, and/or iron rust can be removed only by using other procedures.

Lichen and algae coatings can be removed from most stones by a wire brush, especially from stones that have smooth surfaces and have been completely dried. Do not, however, use a wire brush (hardness ~5.5) on soft rocks such as limestone, dolostone, and soapstone, all of which have hardnesses of less than 5.5. Fortunately, brushes with soft, natural or plastic bristles or directed compressed air will remove many of these coatings from these relatively soft stones.

Some lichen and algae have to be removed by chemical means. A few can be removed by soaking the stones in dilute bases such as ammonium hydroxide ($NH_3OH : H_2O = 20 : 80$). Others, and also calcite coatings, can be removed by soaking the stone in either dilute hydrochloric acid ($HCl : H_2O = 10 : 90$) or vinegar. [Hydrochloric acid (HCl) is sold as muriatic acid in most drug stores.] Do not, however, use acid on stones that contain carbonate minerals—for example, limestones, dolostones, and marbles; the common carbonates are decomposed by these acids.

Rust can be removed from some rocks by merely brushing or airbrasing them. Chemical treatment will be required for others. Soaking the stone in oxalic acid may work. Several other alternative, more complicated procedures have been suggested, but I have found none that works any better than oxalic acid. In any case, if the treatment does not give the results you desire, it is sometimes well worth repeating the treatment; repetition has, in some cases, led to success.

If you want to see how a stone will appear when it is polished or coated, wet it; essentially any relatively clear liquid will work. [For years, I have frequently licked stones, if no other liquid was readily available, in order to get the effect; this is *not* a good thing to do—a few rare stones contain minerals that have sickening tastes or are poisonous.]

So far as giving stones a polished appearance without taking them through the time-consuming tumbling procedure: A coat of wax, shellac, varnish, colorless fingernail polish, or plastic, may be applied as a substitute for polishing. Do not forget, however, that most such coatings are rather easily scratched. Very few stones have to be coated or polished in order to appear shiny when put in mounting plastic.

Actual polishing of stones is usually done by tumbling. When tumbling, a few rules of thumb are well worth keeping in mind. For example, be sure to

Use stones of *nonuniform* size—a feed consisting of mixed sizes will give the best results.

Group the feed stones so that all stones within any given tumbler load have about the same hardness.

Do not expect a uniformly smooth surface to develop on a stone if the major mineral constituents have hardness differences of 2 or more on Mohs' scale.

Do not include flaky, granular, porous, or pitted rocks—very few of them will take on a good polish.

Remember that the overall shapes and sizes of most input stones will be reflected in the the tumbled output—that is to say, if you are interested in making jewelry from your stones, remember that extremely irregular shapes have few uses.

If you wish to take a picture of a stone, photograph it under water. Unpolished stones will have their colors enhanced; polished stones will no longer give undesirable reflections and unwanted highlights.

STORING AND DISPLAYING

Storing a collection can range from keeping a box of stones in a drawer or closet to housing them in permanent custom-made cases. The kind of storage you choose will, of course, depend primarily on your desires and how much time and/or money you are willing to spend.

The way you display your collection will depend on

the type of collection you have and the amount of time and energy you spend in preparing the display. The easiest, and in some ways the best, way to exhibit your collection is to keep each specimen, along with its label, in its own tray or box. This method affords versatility in that the individual trays or boxes can be arranged any way you want them, and you can rearrange the trays any time you wish to change the apparent character or emphasis of your collection.

KINDS OF COLLECTIONS

Every stone collector has his or her own special reason for collecting. Some people collect stones for their colors or their shapes; others collect them just for the joy of having something that has been formed in nature, something that attracted their attention and they could pick up in the great outdoors; the list could go on and on. Indeed, collectors have made many diverse kinds of stone collections ranging from those that are general to those that are highly specialized.

For most novice collectors, a general collection is the best way to start. Two common alternatives are stones the shapes of which are attributable to certain of the diverse processes, mentioned in Chapter 2, and stones composed of one or more kinds of rocks and/or minerals.

The former collection would include stones of various shapes gained as the result of diverse agents of weathering and erosion—for example, stones shaped during stream transport, stones planed and grooved as they were dragged over bedrock by a glacier, stones rounded and smoothed as a result of their to-and-fro movement within the surf zone of some large body of water, and stones abraded by windblown sand and dust. It also might include stones with inherent shapes—for example, volcanic bombs, tektites, and diverse concretions.

The latter collection might include stones made up of fairly resistant minerals (e.g., gold and copper nuggets, ruby and sapphire pebbles, and quartz and agate cobbles) and/or stones made up of one or more of the groups of rocks listed in Table 5.1.

An additional, fairly general type of collection—one that would be of special interest to anyone who enjoys

Table 5.1 Rocks and Pseudo-rocks That Might Be Included in a General Collection

Igneous and Pyroclastic	*Sedimentary and Diagenetic*	*Metamorphic and Migmatitic*	*Others (exhibiting the following:)*	*Pseudo-rocks*
Granite	Conglomerate	Gneiss	Vein	Brick
Syenite	Breccia	Augen gneiss	Dike	Tile
Diorite	Sandstone	Amphibolite	Fault	Glass
Basalt	Arkose	Slate	Fold	Slag
Porphyry	Fossiliferous rock	Quartzite	Stylolite	Coke
Amygdaloidal rock	Limestone	Greenstone	Vug	Cinder
Tuff	Dolostone	Migmatite	Liesegang banding	Porcelain
	Chert	Marble*		Concrete

*Uncommon as a stone.

the art of the lapidary—would consist of stones that are amenable to cutting and polishing, or even to tumbling, and thence fashioning into decorative pieces such as bookends and penstands or even into jewelry. A list of rocks, (many of which are commonly found as stones) that have been used as the raw material for decorative pieces or jewelry is provided in Table 5.2.

Three of the more interesting specialty stone collections that I have seen are

1. A group of between 3,000 and 4,000 ventifacts from some forty worldwide localities. Each of the specimens is marked on the bottom to indicate its compass orientation at its collecting site; as might be suspected, this collection has become the basis of some fundamental wind erosion and aerodynamic studies.
2. A collection of gemstones found as beachstones in the Lake Superior district. This aesthetically oriented collection includes several fine banded agates, diversely colored datolite pebbles, small pebbles of chlorastrolite (Michigan's official State Gem), and pink and white thomsonite "eyes," many of which now have one or more polished surfaces.
3. An intriguing group of stones, each resembling some real or fictitious person, animal, bird, fish, plant, emblem, artifact, geographic entity and so forth (see Figure 5.5). While looking at the curious imitative shapes, now called mimetoliths, that are included in this collection, it becomes easy to understand how ancient humans were led to make up and believe in all sorts of legends relating to stones.

Possibilities for the makeup of specialty collections are, of course, limited only by the imagination and predilection of the individual collector. If any of you knows of an outstanding or markedly different collection of stones, I will greatly appreciate your writing to me about it.

TRADING AND SELLING

After you have collected for a while, you will probably accumulate some specimens that you no longer want. When

Table 5.2 Rocks Commonly Constituting Stones Amenable to Cutting and Polishing

Igneous	*Sedimentary*	*Metamorphic*	*Others*
Basalt (especially amygdaloidal)	Alabaster	Amphibolite	Breccia
"Granite"*	Catlinite	Epidosite	Jasper
Graphic granite	Chert (e.g., flint)	Marble	Migmatite
Obsidian	Conglomerate	Quartzite	Tektite
Porphyry	"Marble"†	Serpentinite	
Rhyolite (especially banded)	Travertine	Soapstone	
	Oolitic and fossiliferous rocks‡	Unakite	
	Organic rocks (e.g., amber, jet, and cannel coal)		

Note: Most of these rocks, briefly defined in the glossary, are described in books about rocks, such as those listed in the Suggested Readings.

*Quotation marks indicate use of this term to include all quartz-bearing phanerites.

†Quotation marks indicate use of this misnomer to refer to limestones and dolostones that take a polish.

‡Includes such well-known stones as the Petoskey stone.

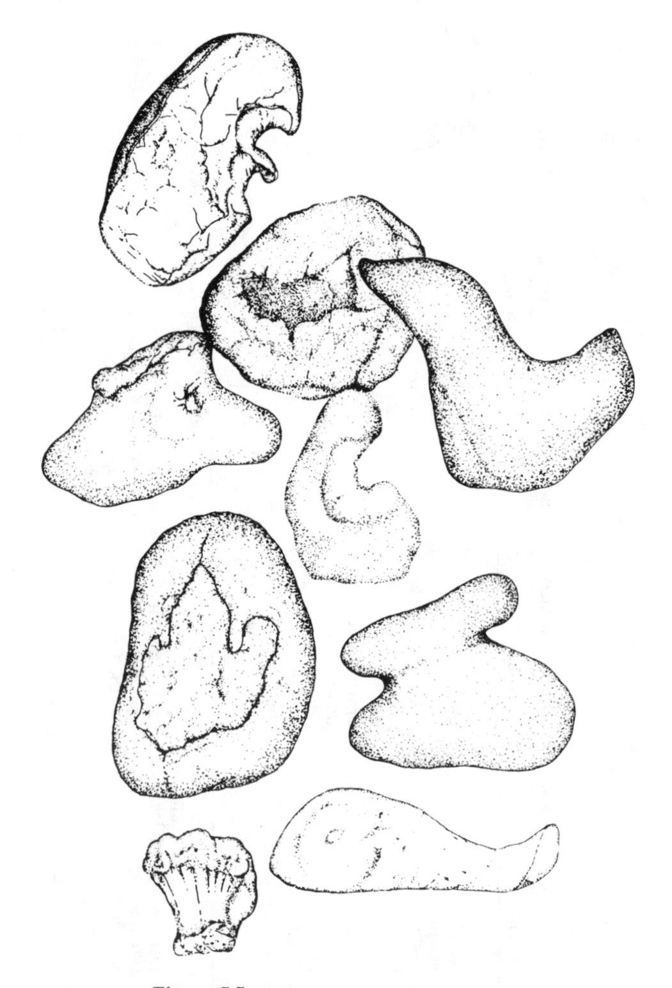

Figure 5.5
Stones with imitative shapes are called *mimetoliths*.

that happens, you should consider selling or trading some of your specimens.

You may find these activities to be enjoyable, just as you do collecting. On the other hand, you may find it traumatic to part with even an "unwanted specimen" that you have collected. Let your feelings be your guide.

If you do decide to trade, buy, and/or sell:

Many clubs hold shows where collectors can trade, sell, or buy specimens. Indeed, this is a stated function of some clubs.

Sometimes, you will be able to get something you really want by trading. You should remember, however, that in order to get a good specimen you will frequently have to give up an equally good specimen.

You may prefer to buy certain specimens. If you do, remember that, just as with any purchase, "let the buyer beware." Also, always remember that the value of any given stone is strictly dependent upon how much someone (including you!) wants it—that is, a stone's value is by no measure absolute. This is even true of most gemstones.

FINDING HELP

As you collect, if you do not already know, you will probably be learning how to identify minerals and rocks and fossils and also how to recognize the features that characterize the different modes of origins of stones. The sketches in Chapters 2 and 3 and the tables in the Appendixes will help you in these endeavors. Nonetheless, whether you are just starting or have been collecting for years, you will almost certainly now and then find a stone the identity and/or history of formation of which baffles you.

When you find yourself on the horns of such a dilemma, your best bet is to turn to an experienced collector or a professional geologist for help. As already mentioned, in many areas, there are museums or nature centers and/or schools or universities in which geology is taught and thus where you can find someone with the appropriate knowledge to help you. Seek out one or more of those people. You will, I feel sure, find someone who not only can but will be quite willing to help you. Indeed, in many cases, (s)he will provide other help, if you need it. For example, (s)he may tell you about additional places where you could go to extend your collecting activities.

When you want to talk to one of these people, it is generally wise to call ahead to make an appointment. This is true because you may not be able to see the person if you drop in unannounced. Also, in some cases, you may be able to find out what you want to know over the telephone.

GOOD LUCK!!!

Appendixes

Mineral Identification

Most of the common minerals can be identified rather easily in hand specimen because of their appearances or by subjecting them to simple tests. The tables given in this appendix will help you to identify the common rock-forming minerals and a few other minerals likely to be seen in stones.

On the tabulation, the minerals are divided into two main groups: those that appear metallic (Table A.1) and those that appear nonmetallic (Table A.2). The minerals in each of these main groups are arranged in order of Mohs' hardness scale and are color coded.

The hardness scale, established by the German mineralogist Friedrich Mohs in the early 1800s, is based on ten fairly common minerals that range from talc, the softest, to diamond, the hardest. The minerals were selected so that each mineral will scratch the minerals that are lower on the scale but will be scratched by minerals that are higher on the scale. The ten minerals that compose the Mohs hardness scale are given in order of increasing hardness on the following page.

Hardness	Mineral	
1	talc	
2	gypsum	
		fingernail
3	calcite	
4	fluorite	
5	apatite	
		knife blade and glass
6	feldspar (orthoclase or microcline)	
7	quartz	
8	topaz	
9	corundum (e.g., ruby and sapphire)	
10	diamond	

As noted on the scale, fingernails have a hardness of about 2.5 and most knife blades and pieces of common window glass have hardnesses of about 5.5. In addition, most high quality metal files have a hardness of about 6.5.

Hardness testing sets may be purchased from many mineral dealers and hobby shops. In fact, hardness pencils—pencillike rods with sharp chips of minerals of known hardnesses mounted on their ends—may be bought, generally in small, easily portable cases. In most instances, however, determining that a mineral is soft (< 2.5), hard (> 5.5) or intermediate (between 2.5 and 5.5) is all that is needed. As an example of how the scale may be used, a mineral with a hardness of 5.5 would scratch apatite, would be scratched by potassium feldspar, and would neither scratch nor be scratched by a knife blade or common window glass.

On the tables, each mineral name is followed by a simple and generally reliable test that will help pin down its identification. Technical terms used on the tables are defined in the Glossary.

The following is a suggested procedure for using the tables:

1. Note whether the mineral is metallic or nonmetallic.
2. Note its color.
3. Determine its approximate hardness.
4. On the basis of entries in the Remarks column, determine other properties that may be definitive.

For example, if the mineral is (1) nonmetallic, (2) white, and (3) has a hardness slightly greater than one's fingernail, it is likely to be either barite or calcite. Subsequently, if it is found to be (4) relatively heavy (compare it with other light-colored minerals and rocks of the same general size), and if it does not effervesce with cold dilute hydrochloric acid, it is probably barite.

One caution: It must be remembered that there is the possibility that the mineral you have found is not included on these tables. If so, perhaps you could identify the unknown mineral by using more extensive determinative tables; for example, those in Chapter 3 of Dietrich and Skinner, 1979 or in Dietrich, 1969. If, however, definite identification of some apparently uncommon mineral is extremely important, it would be wise to have a professional mineralogist or petrologist check the mineral for you.

Table A.1 Mineral Determinative Table for Metallic Minerals (Including Common Submetallic Minerals)

Black/Dark gray	*Yellow/Brassy*	*Red/Brown*	*Hardness*	*Name*	*Remarks*
X			1	Graphite	Submetallic; cleaves into plates; marks on paper.
		X	2½–3	Copper	Branching crystals; malleable and ductile; soluble in HNO_3.
	X		3½–4	Chalcopyrite	Brassy; iridescent tarnish is common.
		X	3½–4½	Pyrrhotite	Bronzy; typically tarnished; commonly magnetic.
X		X	4–5½	Limonite	Submetallic; yellow-brown when powdered.
X		X	5–6	Hematite	Submetallic to earthy; platy to granular; red-brown when powdered.
X			5½–6½	Magnetite	Octahedral crystals or granular; strongly magnetic.
	X		6–6½	Pyrite	Brassy; cubes and 12-sided crystals are common.

Table A.2 Mineral Determinative Table for Nonmetallic Minerals

Colorless/White	*Gray/Black*	*Red/Violet*	*Green/Blue*	*Yellow/Brown*	*Hardness*	*Name*	*Remarks*
X	X		X		1	Talc	Feels soapy.
X	X			X	1–3	Clay	Earthy to pearly; sticks to tongue; smells clayey when breathed on (group name for several hydrous aluminum silicates).
X					2	Gypsum	Pearly to massive; soluble in HCl.
			X		2–3	Chlorite	Pearly; cleaves into thin inelastic plates.
X			X		2–4	Mica: muscovite	Cleaves into thin elastic plates.
	X					biotite	
				X		phlogopite	
	X		X	X	2½–4	Serpentine	Waxy to greasy; harder than talc.

Table A.2 Continued

Colorless/White	*Gray/Black*	*Red/Violet*	*Green/Blue*	*Yellow/Brown*	*Hardness*	*Name*	*Remarks*
X					3	Barite	Heavy for a white mineral; some is fetid when rubbed.
X	X			X	3	Calcite	Effervesces briskly in dilute HCl; cleaves into blocks with diamond-shaped faces.
X	X				3½	Dolomite	Effervesces slowly with dilute HCl only when powdered.
X		X		X	4	Fluorite	Cubic crystals are common; cleaves into eight-sided blocks; some fluoresces in ultraviolet light.
	X			X	4–5½	Limonite	Yellow-brown when powdered (field name for hydrous iron oxides).

	X	X		X	5–6	Hematite	Submetallic to earthy; red-brown when powdered.
	X		X		5–7	Amphibole	Two cleavages at 55° and 125°; lathlike.
	X		X		5–7	Pyroxene	Two cleavages at 90°; blocklike.
			X		6	Epidote	Pistachio to bilious green.
X		X			6–6½	Potassium feldspar	Two cleavages at 90°; flesh to salmon color is common.
X	X				6–6½	Plagioclase feldspar	Two cleavages at nearly right angles; parallel lines may be seen on better cleavage.
		X		X	6–7½	Garnet	Glassy to dull; equal-sided crystals; irregular fracture.
		X		X	6½	Chalcedony (agate, etc.)	Mammillary; microcrystalline; waxy luster; agate is color-banded; carnelian is reddish brown; sard is light to dark brown.
			X		6½–7	Olivine	Glassy; apple green; conchoidal fracture.
X	X			X	7	Quartz	Glassy; conchoidal fracture.
	X			X	7	Tourmaline	Glassy; lengthwise striated prisms; commonly has cross sections that resemble spherical triangles.
	X			X	7½	Staurolite	Cross-shaped crystals are common; fires to magnetic powder.

B

Animals and Plants Common in the Fossil Record

A **fossil** *is a life form or evidence of life recorded in rock.*

To examine the different aspects of this definition: *Life form* covers any complete or distinguishable part of an animal or a plant. *Evidence of life* is generally considered to refer to such things as tracks, burrows, coprolites (petrified excrement), and calculi (gallstones, kidney stones, and so forth). A few scientists also include structures (artifacts) produced by prehistoric man. *In rock* serves primarily to emphasize the antiquity of fossils and the requirement of natural preservation.

It is clear that, in origin, fossils are biological as well as geological. Therefore, in naming fossils it is standard procedure to use the same binomial system of nomenclature that biologists use. This dual name consists of genus and species designations. When written, both names are italicized and the generic name is capitalized and precedes the species name that is not capitalized (for example, *Homo sapiens*). In some cases, the last name of one or more persons is given after the species name; this indicates that the species was originally described by the person or persons named.

The complete biological classification scheme is based on kinship and similarities of anatomical features. Three examples—one for a vertebrate, one for an invertebrate, and one for a plant—are given in Table B.1 Resemblances among differently named members of each hierarchy increase from top to bottom. A diagram of the scheme would be similar to that of a tree with its trunk called phylum; its main branches, classes; its smaller branches (in descending order of size), orders, families, and genera; and the twigs, species.

The following elaboration should help you to understand the classification scheme. It is based on the first example given in Table B.1.

The Kingdom Animalia is generally considered to include all life except plants.

The Phylum Chordata includes all animals with gill pouches, nerve chord, and notochords.

The Subphylum Vertebrata (not given on the table) requires the presence of a backbone.

The Class Mammalia adds such requirements as warm blood, mammary glands, and hair.

The Order Primata includes only animals with superior nervous systems, opposing thumb-finger relationship, and nails on digits.

The Suborder, or superfamily, Anthropoidea (not given in the table) includes apes, humans, and monkeys.

The Family Hominidae includes our close relatives, who have bipedal posture, nonopposable big toes, legs longer than arms, no tail, and certain other characteristics.

The Genus *Homo* includes anthropoids.*

The Species *sapiens* (of *Homo*) is modern man—contrasted to, for example, *H. neanderthalensis*, one of the prehistoric species of *Homo*.

The different races—australoid, caucasoid, mongoloid, and negroid—would be subspecies.

Species are distinct in that their members can only breed with others of their kind and thereby produce offspring that have the same capability. For extinct species, known only in the fossil record, it cannot be known that such breeding restrictions prevailed. Therefore, for these

---

*With humans, generic and specific definitions may be biologically or anthropologically based. Biological criteria depend on sizes and shapes of anatomical features; anthropological criteria are based on such things as tool-making and social parameters that measure intangible characteristics like intelligence.

species morphological appearances, geographic or stratigraphic distributions, and indicated behavioral characteristics are considered definitive.

Animal phyla known to be well represented in the fossil record are listed in Table B.2. Well-represented plant phyla are given in Table B.3. Descriptions and pictures of representatives of the different phyla may be found in many elementary biology and paleontology textbooks.

It is noteworthy that in stones, many fossils are represented only by fragments (see, for example, Figure 3.11D). In many cases, however, the fragments are sufficient to identify the life form from which they came.

Table B.1 Biological Classification

	Vertebrate	*Invertebrate*	*Plant*
Kingdom	Animalia (also Metazoa)	Animalia (also Metazoa)	Plantae (also Metaphyta) Embryophyta (subkingdom)
Phylum (Division in botany)	Chordata	Coelenterata	Tracheophyta Pterophytina (subdivision)
Class	Mammalia	Anthozoa Zoantheria (subclass)	Angiospermopsida Monocotylendonae (subclass)
Order	Primata	Rugosa (Tetracoralla)	Typhales
Family	Hominidae	Disphyllidae	Typhaceae
Genus	*Homo*	*Hexagonaria*	*Typha*
Species* (common name)	*H. sapiens* (human)	*H. percarinata* (the Petoskey stone, Michigan's state stone)	*T. latifolia* (the common cattail)

*Species names alone are meaningless in that species of different genera may be given identical names.

Table B.2 Animals Well Represented in the Fossil Record

Phylum *Subphylum*	*Class* *Subclass*	*Common Names or Description*
Protozoa	Sarcodina	Foraminifera, radiolaria
Porifera		Sponges
Coelenterata	Hydrozoa	Hydroids, *stromatoporoids*,*
	Anthozoa	Sea anemones, *tetracorals*, hexa-corals, *tabulate corals*
Brachiopoda		Brachiopods
Bryozoa (or Ectoprocta)		"Moss corals"
Echinodermata	Stelleroidea	Starfish, brittle stars
	Echinoidea	Sea urchins, sand dollars
	Crinoidea	Sea lilies
	Blastoidea	*Blastoids* (or "*sea buds*")
	Cystoidea	*Cystoids*
Mollusca	Pelecypoda (or Lamellibranchiata)	Clams, oysters, scallops
	Gastropoda	Snails, conchs, etc.
	Cephalopoda	Squids, nautiloids, *ammonoids, belemnoids*

Annelida		Earthworms, etc. (segmented worms)
Arthropoda	Trilobita	*Trilobites*
	Crustacea	Lobsters, barnacles, crabs, ostracodes
	Arachnoidea	
	Arachnida	Spiders, ticks, scorpions
	Merostomata	*Eurypterids*, king (or horseshoe) crabs
	Insecta	Insects
Chordata		
Vertebrata	Pisces	Fishes
	Amphibia	Salamanders, frogs, toads, *stegocephalians*
	Reptilia	Crocodiles, lizards, turtles, *dinosaurs, ichthyosaurs, plesiosaurs, mosasaurs, pterodactyls*
	Aves	Birds
	Mammalia	Horses, cattle, elephants, whales, humans, kangaroos, bats, cats, rats, etc.
Hemichordata	Graptolithina	*Graptolites*
Unknown	Unknown	Conodonts

SOURCE: Reprinted by permission, from R. V. Dietrich, *Geology and Virginia*, Copyright © 1970 University Press of Virginia.

*Names of extinct creatures are italicized.

Table B.3 Plants Well Represented in the Fossil Record

Phylum	*Class*	*Common Names or Description*
	I. Subkingdom Thallophyta	
Cyanophyta		Blue-green algae (Cryptozoons?)
Chlorophyta		Green algae
Euglenophyta		Euglenoids
Charophyta		Charophytes
Phaeophyta		Brown algae
Rhodophyta		Red algae
Chrysophyta		Golden-brown algae, diatoms
Pyrrophyta		Cryptomonads, dinoflagellates
Myxomycophyta		Slime molds
Eumycophyta		True fungi
	Phycomycetes	Algaelike fungi
	Ascomycetes	Sac fungi
	Basidiomycetes	Club fungi
	Deuteromycetes	Fungi imperfecti
Schizomycophyta		Bacteria

II. Subkingdom Embryophyta		
Bryophyta		
	Hepaticae	Liverworts
	Musci	Mosses
Tracheophyta		Vascular plants
Subphyla:		
Psilopsida		
Lycopsida		Club mosses, *Sigillaria*,* *Lepidodendron*
Sphenopsida		"Horsetails," *Calamites*
Pteropsida		
	Filicineae	True ferns
	Gymnospermae	Ginkgos, seed ferns, conifers, *Cordaites*
	Angiospermae	Flowering plants, palms, grasses

SOURCE: Reprinted, by permission, from R. V. Dietrich, *Geology and Virginia*, Copyright © 1970 The University Press of Virginia.

*Italicized names are for extinct plants.

C

Rock Identification

With a little experience, almost anyone can learn how to identify most of the common rocks. A good starting point is to use the tables and procedure outlined in this appendix.

It must be kept in mind, however, that only the rock types that have been found to constitute fairly common stones are included on these tables. Therefore, to identify any relatively uncommon rocks, you will have to consult some more comprehensive tables—for example, those given in Chapter 8 of Dietrich and Skinner, 1979—or a professional petrologist.

Table C.1 is for rocks the main constituents of which can be distinguished with the naked eye or with the aid of a 10× handlens. Table C.2 is for rocks the constituents of which are too small to be so distinguished.

On each table, there are two main subdivisions based on hardness and there are additional subdivisions based on rock origins. The former are indicated by a heavy horizontal line on each table; the latter are separated into columns. On Table C.1, there are also broader subdivisions that are separated by vertical lines. In addition, there are remarks about each of the rocks or rock groups.

Prophyries, some of which occur fairly commonly as stones, are not included on the tables. To name a porphyry, you should identify the groundmass by using the tables and then name the rock according to the formula *groundmass-name + porphyry*; for example, a rock that is constituted by phenocrysts within a basaltic matrix would be called a *basalt porphyry*.

The two hardness categories on the tables are based on the hardness of a jackknife blade. This basis is used because most collectors carry a knife or have one readily available and most of the common rock constituents are definitely harder or softer than the typical knife blade.

One caution: You should always determine a rock's hardness by using the rock to scratch the knife blade, rather than the knife blade to scratch the rock. If you use this method, you are not likely to be misled by the disaggregation of any of the relatively incoherent rocks. In addition, you should check several parts of each rock so that some unusual constituent grain does not give you a false impression of the general hardness of the overall rock.

The genetic subdivisions are those given in Chapter 3. To facilitate the use of the tables, diagenetic rocks and coals are combined with the sedimentary rocks and pyroclastic rocks are combined with the igneous rocks. The remarks give simple tests or other data that will help you to distinguish one rock from other rocks that it may resemble; they deal with characteristics such as how a rock fractures, its mineral content, and its solubility.

The following are two examples of how the tables may be used:

I. 1. The rock has macroscopically discernible grains: Use Table C.1.
   2. The hardness is less than that of a knife blade: Go to the bottom part of the table.
   3. The texture is interlocking with no obvious preferred orientation: Note the left third of the bottom part of the table.
   4. The rock is one of the following:
      A. limestone or calcite marble if the rock effervesces briskly with dilute HCl;
      B. dolostone or dolomite marble if it effervesces slowly with dilute HCl.

II. 1. The rock consists of grains too small to be seen macroscopically: Use Table C.2.
   2. The hardness is greater than that of a knife blade: Go to the top half of the chart.
   3. The rock is one of the following: felsite, basalt, obsidian, pumice, ash tuff, chert, shale, slate, phyllite, greenstone, or mylonite.
   4. Each of those rocks may be distinguished from the others on the basis of the remarks—for example, a chert could be distinguished from the others because of its luster and fracture.

Good luck!

Table C.1 Identification of Rocks That Have Macroscopically

	Interlocking Grains			
	Igneous	*Sedimentary or Diagenetic*	*Metamorphic*	
←— Hardness > knife blade —→	After constituent minerals are identified, see Table 3.1		*Quartzite*. Vitreous, conchoidal fracture. *Epidosite*. Chiefly green epidote.	
←— Hardness < knife blade —→		*Limestone*. Brisk effervescence with dilute HCl. *Dolostone*. Slow smoldering effervescence with dilute HCl.	*Marble*. May be calcitic or dolomitic; commonly contains disseminated graphite or silicate minerals; effervesces as noted to left.	

Discernible Grains and Are Commonly Found as Stones

Foliated *Metamorphic (interlocking grains common)*	**Fragmental or Layered** *Igneous or Pyroclastic*	*Sedimentary or Diagenetic*	*Metamorphic*
(Quartz is common in all but amphibolite.) *Gneiss*. Streaked or banded; chiefly granular grains. *Amphibolite*. Dark gray to greenish black. *Schist*. Enough platy constituents to give good foliation.	Pyroclastic rocks are named on basis of size of fragments that make up 50 percent or more of their volume. *Dunite*. Looks like an olivine sandstone.	After fragments are identified as to size, see Table 3.2. *Graywacke*. A gray or green sandstone; clay odor when damp. *Arkose*. A sandstone with 25 percent or more feldspar.	*Quartzite*. Vitreous, conchoidal fracture. *Metaconglomerate*. Pebbles, cobbles, and boulders in a meta-quartzite matrix.
Soapstone. Greasy appearing; commonly veined; very soft. *Serpentinite*. Typically green; resembles soapstone but is harder than fingernail.		*Clastic limestone*. Contains fragments such as fossils and oolites; effervesces briskly with dilute HCl. *Dolostone*. As above, but effervesces slowly with dilute HCl.	

Table C.2 Identification of Microcystalline or Glassy Rocks Commonly Found as Stones

	Igneous	*Pyroclastic*	*Sedimentary or Diagenetic*	*Metamorphic*
Hardness > knife blade	*Felsite* (e.g., rhyolite and dacite). Stony appearance; light colored. *Basalt*. Dark gray to greenish black; commonly vesicular or amygdaloidal. *Obsidian*. Glassy; translucent in thin pieces; brown or black. *Pumice*. Glassy; frothlike.	*Ash tuff*. Consolidated volcanic ash; rough feel.	*Chert*. Porcelaneous luster; conchoidal fracture. *Shale*. Is fissile; clay odor when damp; disaggregates with knife rather easily; less compact than slate.	*Slate*. More compact than shale; parallel rock cleavage, commonly at angle to bedding. *Phyllite*. Glossy sheen on foliation surfaces; commonly is corrugated. *Greenstone*. Olive green; dull luster; subconchoidal fracture. *Mylonite*. Smeared-out appearance; commonly includes sheared fragments.
Hardness < knife blade			*Coal*. Black color and streak; brittle. *Limestone*. Brisk effervescence with dilute HCl; chalk variety is crumbly. *Dolostone*. Slow, smoldering effervescence with dilute HCl.	*Mylonite*. As above.

Glossary

Each of the meanings given for terms in this glossary is implied by the word's use in this book. None should be considered either hard-and-fast or all-inclusive. Only the relatively common rocks and rock types are included and no minerals or fossils are given. For additional definitions of geological terms, see the American Geological Institute's *Glossary of Geology* or *Dictionary of Geological Terms* (complete bibliographic data are provided in the General References section of the Suggested Readings).

abrasion mechanical grinding or wearing down by impact of projectiles or by friction.

accessory mineral in an igneous rock, a minor constituent that is not taken into account in classifying and naming the rock.

aeolian (or eolian) relating to natural wind activities or deposits.

aetite synonym for *eagle stone.*

agate color-banded chalcedony (cryptocrystalline quartz).

agglomerate a pyroclastic rock consisting predominantly of volcanic bombs.

alabaster a white or diversely tinted, fine-grained rock made up completely or nearly completely of the mineral gypsum.

alkali feldspar potassium feldspar (microcline or orthoclase), sodium-rich plagioclase (albite), their perthitic combination, or any mixture thereof.

alluvial relating to stream (i.e., running water) activities or deposits.

alluvial fan a stream deposit, roughly the shape of the Greek letter delta (Δ), formed where the stream flows out of a narrow mountain valley onto a rather flat and open area such as an intermontaine basin.

alteration product a mineral component formed from a preexisting mineral as the result of partial chemical change.

amber a relatively hard, brittle, yellowish to brownish, transparent or translucent fossil resin (from coniferous trees).

amphibolite a metamorphic rock, typically well foliated, that consists largely of an amphibole and a plagioclase feldspar. Amphibolites contain negligible quartz.

amygdule (adj. *amygdaloidal*) a filled vesicle.

andesite the aphanitic equivalent of diorite.

anorthosite an igneous phanerite consisting of 90 percent or more plagioclase feldspar, typically bluish-gray labradorite.

anthracite coal with a high fixed carbon content (92–98 percent) and a very low volatile content. It typically has a high luster and conchoidal fracture; it is sometimes called "hard coal."

aphanite crystalline rock in which the individual grains are so small that they cannot be distinguished by the unaided eye or by the eye aided only by a 10 × handlens.

aplite an igneous rock of granitic or granodioritic composition and having a saccharoidal texture.

arkose a sandstone consisting of at least 25 percent feldspar and more or less resembling granite.

artifact object made by humans or humanoids, such as a flint weapon or tool.

ash (volcanic) unconsolidated pyroclastic material made up largely of fragments less than 2 mm (~1⁄10 inch) in intermediate diameter.

atom the smallest component of an element that has all of the properties of the element.

attrition reduction of tools of abrasion caused by impact breakage.

augen gneiss gneiss that consists in part of eye-shaped grains of a mineral, typically a feldspar (see Figure 3.14). (*Augen* means "eyes" in German.)

bajada a broad, gently inclined surface formed by coalesced alluvial fans that extend out from the base of a mountain front.

basalt the aphanitic equivalent of gabbro. Most basalts are dark gray to greenish black in color.

batholith an intrusive igneous mass having no known bottom and a bedrock surface of 100 sq. km (~40 sq mi) or more (see Figure 3.3).

beach shore area that extends from the low water line to the area of change in topography or vegetation.

bedrock the continous solid rock that is exposed at the surface or is directly beneath the overburden.

bituminous coal coal that contains 15 to 20 percent volatiles and thus contains less fixed carbon than anthracite. It is typically banded with high and dull luster laminae; it is the "soft coal" of commerce.

boulder a rounded stone with its intermediate diameter greater than 256 mm (~10 inches).

boulder train a line or only slightly dispersed group of glacial boulders that extends out from a single bedrock source in the direction that the transporting glacier moved.

breccia any rock consisting of angular rock fragments (larger than sand size) in a finer-grained matrix; for example, consolidated rubble or lithified fragments formed as a result of crushing in a fault zone.

calcarenite a limestone consisting of sand-sized fragments.

calcilutite a limestone consisting of silt or clay-sized particles.

calcirudite a limestone consisting of pebble-sized or larger fragments.

cannel coal dull, compact, homogenous coal—of bituminous rank—that tends to exhibit conchoidal fracture. (Under the microscope, it can be seen to consist largely of spores and pollen).

catlinite brownish red claystone frequently used by the Sioux Indians for carving of such things as peace pipes.

cave pearl a calcite concretion, typically pea-sized or smaller, formed within a cavern pool.

chemical weathering the breaking down of a rock as a result of such reactions as oxidation and hydration; syn., decomposition.

chert a diagenetic or nondetrital sedimentary rock composed of microcrystalline quartz.

clastic see *detrital.*

claystone a detrital sedimentary rock consisting largely of clay particles.

cleavage the tendency of a crystalline substance to break or split along plane surfaces the positions of which are controlled by the substance's atomic structure.

coal a natural fuel consisting of solid hydrocarbons.

cobble a stone with its intermediate diameter between 64 and 256 mm (~2½ and 10 inches).

colluvial relating to downslope movement under the influence of gravity and to the materials so moved.

competence (of streams) the size of the largest fragment of a given specific gravity that a stream will move. It is dependent upon the stream's velocity.

conchoidal fracture seashell-like concavity formed when brittle materials such as glass or quartz are broken.

concretion an ellipsoidal or irregularly shaped mass, typically developed within sedimentary rocks by localized deposition around some nucleus (see Figure 2.25).

conglomerate a detrital sedimentary rock that consists predominantly of rounded stones (see Figure 3.9).

contact metamorphism metamorphism genetically related to magma and taking place at or near the contact between a consolidating magma and the surrounding country rock.

country rock the rock into which magma has been intruded.

crystal (as applied to tuffs) consisting of mineral fragments; cf. *lithic* and *vitric.*

crystal structure the periodic repetition of any particular arrangement of atoms (see Figure 1.2).

dacite the aphanitic equivalent of granodiorite.

decomposition see *chemical weathering*.

dendrite a mineral growth, typically of manganese oxide, that branches irregularly like a tree (see Figure 3.17).

desert varnish a lustrous coating of iron-oxide–rich mineral matter, typically brown or black in color, that occurs on stones and other exposed rock surfaces, generally in desert regions. It is thought to form by evaporation of mineral-bearing solutions exuded or dissolved from the rock material that is coated.

detrital consisting of naturally broken-off pieces of preexisting rocks. The term is applied to both sediments and sedimentary rocks made up of such fragments; syn., *clastic*.

diabase a fine-grained gabbro.

diagenetic pertaining to changes—such as recrystallization and replacement—that take place in sediments after their deposition but before, and commonly contributing to, their conversion to rock; also applied to rocks consisting largely of materials formed in response to such changes.

differential erosion processes whereby different parts of nonuniform materials react differently in response to agents of erosion.

differential pressure forces that are not equal on different sides of a body—that is, forces that tend to or do cause distortion.

differential weathering processes whereby different parts of nonuniform materials have reacted differently in response to disintegration or decomposition.

dike a tabular igneous mass that transects the bedding or foliation of the surrounding country rock (see Figure 3.3).

diorite an igneous phanerite typically consisting of 50 to 70 percent of a light-colored plagioclase, less than 5 percent quartz, and 25 to 50 percent varietal minerals (hornblende $\pm$ biotite $\pm$ augite).

disintegration see *physical weathering*.

dolostone a diagenetic or nondetrital sedimentary rock composed largely of the mineral dolomite.

drift overall name for glacially deposited materials.

drumlin a deposit of glacial drift with the topographic expression of a smoothly rounded hill or ridge having the overall shape of an inverted spoonbowl. The long axis is roughly parallel to the direction of ice movement with the blunt end pointing in the direction from which the ice approached (see Figure 2.10).

druse an aggregate of minute crystals, typically covering the walls of a vug or other cavity in rock.

dynamic metamorphism metamorphism caused largely, if not exclusively, by deformation (for example, by faulting).

eagle stone a clay-ironstone concretion with a loose stone in its interior, once believed to have been taken by eagles to their nests to help their mates to produce young.

effervesce to bubble or hiss as, for example, when dilute hydrochloric acid is put on calcite or limestone.

eolian see *aeolian*.

epidosite an epidote-quartz rock, generally with a granular texture, of metamorphic origin.

erosion the weathering and transportation of rock or unconsolidated rock materials.

erratic see *glacial erratic*.

esker a relatively long, commonly sinuous ridge made up of stratified sand and gravel deposited by an englacial or subglacial stream and left behind when the ice melted. The stratification of "let down" eskers is generally somewhat disrupted (see Figure 2.10).

extrusive relating to magma ejected out onto the earth's surface; also applied to the volcanic rocks formed from such magma.

fault a rock fracture with a lateral displacement of the rocks on one side of the break with respect to those on the other side (see Figure 3.22). The locus of fracture, typically a zone up to several meters thick, is generally called a *fault zone*.

fissile said of rocks that split along fairly closely spaced planes parallel to stratification laminae.

fixed carbon content in coal, the solid material, other than ashes, that would remain if the coal were destructively distilled.

flagstone said of slabs produced by splitting rocks along bedding, foliation, or rock cleavage. Such stones are widely used for patios, walkways, and so on.

flint the name frequently given to chert, especially that used for artifacts.

fluoresce to emit visible light when exposed to ultraviolet ("black") light.

fluvial relating to stream (i.e., running water) activities or deposits; syn., alluvial.

fold a bend in rock strata or foliation (see Figure 3.21).

foliation the streaked or banded appearance of a metamorphic rock that is the manifestation of the preferred orientation of platy or rodlike constituent mineral grains (see Figure 3.13).

fossil a life form or evidence of life recorded in rock.

fragmental see *detrital*.

frost heaving moving, generally lifting, of loose material as a consequence of the freezing of water. (Water expands by almost 10 percent when it is frozen to ice.)

gabbro an igneous phanerite typically consisting of 50 to 90 percent dark-colored plagioclase, less than 5 percent quartz, and 35 to 65 percent varietal minerals (augite ± another pyroxene ± olivine).

gastrolith a stone formed or ingested and used in the digestive tract of a vertebrate—for example, a gizzard stone.

genus (plural *genera*) a group of closely related species, possibly descendants of a common ancestor.

geode a spheroidal or ellipsoidal, partly to completely crystal-filled nodule that occurs in some limestones, dolostones, and calcareous shales; geodes' border zones typically differ in composition from and thus are easily separable from their host rocks.

geology the study of the earth. Geology includes the investigation of earth materials—minerals (mineralogy) and rocks (petrology), including fossils (paleontology); earth features—land forms (geomorphology), layered formations (stratigraphy), and disrupted rocks (structural geology); and the application of these studies to help fulfill human needs and desires—economic geology, petroleum geology, engineering geology, and environmental geology. Geology also is fundamental to such interdisciplinary subjects as geophysics, geochemistry, and biogeochemistry.

glacial pertaining to both the activities of the deposits by glacial ice (see Figure 2.10).

glacial erratic a glacially transported and deposited stone, generally of boulder size, that is of different lithology or age than the underlying bedrock.

glacio-fluvial pertaining to activities of and deposits by glacial meltwater streams; for example, eskers and outwash.

glass a liquid with so high a viscosity that it appears solid. (The constituent atoms and ions do not have a regular arrangement like those in crystalline materials.)

gneiss a roughly foliated or banded metamorphic rock consisting largely of granular minerals such as quartz (see Figure 3.13).

gradient (of a stream) the measure of the grade (fall) of a stream. It is usually expressed in meters per kilometer (or feet per mile).

granite an igneous phanerite consisting of 20 percent or more quartz, more alkali feldspar than plagioclase, and between 5 and 20 percent varietal minerals (typically biotite ± muscovite ± hornblende).

granodiorite an igneous phanerite consisting of 20 percent or more quartz, at least twice as much plagioclase as alkali feldspar, and 5 to 25 percent varietal minerals (typically hornblende + biotite.)

granulite a nonfoliated metamorphic rock consisting completely, or almost completely of granular minerals. Because of inconsistent use by different geologists, the term should be abandoned as a rock name except for preliminary field use.

graphic granite a granite having a texture in which quartz within potassium feldspar resembles cuneiform characters (Figure 3.7).

gravel a detrital sediment consisting of 50 percent or more by volume of rounded fragments greater than 2 mm (~1⁄10 inch) in intermediate diameter.

graywacke an impure, typically gray or green sandstone consisting of 25 percent or more dark-colored mineral or rock fragments.

greenstone a field designation for metamorphosed basalts that appear green because of the presence of chlorite, epidote, or a green amphibole.

groundmass the matrix that surrounds the phenocrysts of a porphyry (see Figure 3.5).

groundwater subsurface water. Some geologists restrict this term to water in the saturated zone below the groundwater table.

hornfels a rock formed as the result of pure thermal metamorphism within a contact zone around a magma. Hornfelses are typically dense and nonfoliated.

hydration the chemical combination of water with a substance to form a new substance.

hydrocarbon a compound of hydrogen and carbon.

hydrostatic pressure used to describe a force field in which all pressures are equal; that is, there is uniform stress from all directions.

igneous said of rock formed by the cooling and consolidation of magma; also said of processes directly involving magma.

inorganic containing no free carbon of organic origin.

intrusive said of igneous masses consolidated beneath the surface of the earth.

ion an atom that has lost or gained one or more electrons and, as a consequence, has become negatively or positively charged.

iridescence exhibiting multicolored reflections; for example, the play of colors given by films of oil on water.

isostatic rebound upward recovery of the earth's crust in response to the removal of a load that was great enough to cause crustal depression.

jasper a red or brownish red chert.

jet a relatively hard, highly lustrous coaly material most of which occurs as isolated masses here and there in some carbon-rich shales.

joint rock fracture with no displacement except that

essentially perpendicular to the break—that is, with no lateral displacement of the rocks on one side with respect to those on the other side of the fracture.

kame a deposit of poorly- to well-sorted and stratified glacio-fluvial sand and gravel that was deposited in contact with glacial ice. Many kames are of uncertain mode of formation.

laccolith a floored, typically lens-shaped, igneous intrusive (see Figure 3.3).

lacustrine relating to lake deposits and activities.

lag gravel a residual deposit of stones formed as a result of the winnowing out or blowing out and removal of the previously admixed finer particles.

laminae (singular *lamina*) a thin layer of sediment that is typically 0.05 to 1.0 mm thick.

lapilli volcanic fragments of pebble and cobble size (2 –256 mm, ~1⁄10 to ~10 inches).

larvikite a syenite made up largely of light to dark bluish gray feldspars that exhibit an opalescence.

lava magma that has been extruded onto the earth's surface.

liesegang banding color-banding, typically nested (so as to appear concentric on flat surfaces), that is widely thought to have been formed by some sort of rhythmic precipitation in a fluid-saturated rock (see Figure 3.16).

limestone a sedimentary rock made up largely or wholly of the mineral calcite.

lithic (as applied to tuffs) consisting of rock fragments; cf. *crystal* and *vitric*.

lithified converted to rock.

lithological relating to the petrographic makeup of a rock.

luster the appearance of a surface in reflected light.

magma molten or partially molten rock material. To be called magma, that which is only partially molten must be at least potentially mobile.

mammillary said of mineral surfaces that resemble clusters of partial spheres.

manganese nodule a brown or black concretion, typically of about golf ball size, constituted chiefly of manganese oxide minerals (in some nodules, alternating with layers rich in iron oxides) and occurring in large numbers here and there on ocean floors.

marble a metamorphic rock made up largely of calcite or dolomite.

marine pertaining to oceans and seas, their activities and deposits.

mass wasting the downslope movement of rock material under the influence of gravity.

meta-igneous short for *metamorphosed igneous*.

metamorphic describes rock formed by the transformation of preexisting rock in response to elevated temperature or pressure, sometimes accompanied by changes in chemical environment (for example, in the composition of percolating fluids). The processes are termed *metamorphism*.

metasomatism metamorphism caused chiefly by the introduction of fluids or ions.

meteorite a stone that has fallen to earth from outer space.

microcrystalline consisting of crystalline material distinguishable only under a high-powered microscope.

micromount a small mineral specimen (~ 0.5–1.0 cm, or ~ 3⁄16–3⁄8 inches), typically exhibiting well-developed crystal faces, that is generally viewed through a binocular microscope.

migmatite a macroscopically composite rock typically made up of a dark-colored amphibolite or biotite gneiss intimately mixed with a light-colored rock of granitic to granodioritic composition (see Figure 3.18).

mineral a natural substance, generally inorganic, with a characteristic internal arrangement of atoms and a chemical composition and physical properties that either are fixed or vary within a definite range.

moraine any accumulation of glacial till deposited either directly by a glacier or by being "let down" as the glacier melted.

mudstone a detrital rock consisting of clay-sized particles and lacking fissility; cf. *shale*.

mylonite a rock formed by dynamic metamorphism—that is, by extreme microbrecciation within a fault zone.

obsidian an igneous rock made up of dense glass with a silica (SiO_2) content greater than about 60 percent.

oolite an ellipsoidal mass, resembling fish roe, with internal concentric or radial structures. (Oolites are commonly composed of calcium carbonate—calcite or aragonite—but also may be composed of other minerals such as chalcedonic quartz.)

outcrop a natural exposure of bedrock.

outwash stratified sand and gravel washed out and deposited beyond the margin of the source glacier.

overburden any unconsolidated natural rock material overlying bedrock.

oxidation the chemical combination of oxygen with another element or group of elements.

patterned ground term applied to well-defined patterns, typically outlined by concentrations of stones, thought to have formed in response to frost action within overburden above permafrost.

pebble a stone with its intermediate diameter between 2 and 64 mm (~1/10 and 2½ inches).

pediment a gently inclined, fairly flat erosion surface that is typically more or less covered by a veneer of alluvial gravel.

pegmatite any exceptionally coarse-grained igneous rock or vein that consists largely of quartz, alkali feldspar, and mica.

Pele's hair threadlike strands of volcanic glass.

Pele's tears droplike masses of volcanic glass, many of which have trailing Pele's hair.

percussion marks crescentic cracks on rocks, most commonly rocks that are fine grained—for example, a quartzite.

peridotite an ultrabasic igneous phanerite consisting of at least 50 percent olivine.

permafrost permanently frozen ground—that is, overburden or bedrock that has been constantly frozen for any period longer than twelve months.

petrology the science dealing with rocks, their nomenclature, classification, and origin.

phanerite an igneous rock in which the specific minerals may be distinguished with the naked eye or under simple handlens.

phenocryst a relatively large grain, commonly with crystal outlines, surrounded by a finer grained matrix (groundmass) in a porphyry (see Figure 3.5).

phyllite a foliated metamorphic rock consisting of very fine crystalline grains that give the rock a glossy sheen. Many phyllites appear corrugated.

physical weathering the breaking down of rock as the result of loosening and breaking up of large fragments into smaller ones without changing the composition of the original substances; syn., *disintegration*.

placer a surficial deposit of fragments of heavy minerals usually formed by mechanical concentration in streams but, in some cases, formed by aeolian, lacustrine, or marine processes.

porphyry an igneous rock in which relatively large and conspicuous grains (phenocrysts) are surrounded by a finer-grained matrix, generally called the groundmass (see Figure 3.5).

pothole a hole worn into bedrock by stones being spun around and around in an eddy current (see Figure 2.8).

pseudo-rock any material that resembles rock in both appearance and properties—for example, brick, cinders, concrete, and slag—but is not natural.

puddingstone the popular name frequently applied to conglomeratic rocks whose pebbles have colors that contrast markedly with the color of the matrix.

pumice a frothlike volcanic rock of obsidian composition.

pyroclastic rock consolidated volcanic fragments that were extruded explosively and deposited by settling (like sediment) either on land or in water.

quartzite a metamorphic or sedimentary rock consisting almost wholly of macrocrystalline quartz.

recrystallization formation of new crystalline grains where others once existed; for example, the formation of large interlocking grains of calcite where there were fine particles of calcite or aragonite.

regional metamorphism metamorphism of rocks underlying an area of several hundreds of square kilometers.

replacement any chemical process thought to involve essentially simultaneous removal of one mineral and deposition of another in its place.

residual boulder a boulder formed as the result of weathering, usually chemical weathering. Such boulders commonly exhibit an onionshell-like appearance (see Figure 2.14).

residuum unconsolidated rock material formed essentially where it now occurs—that is, the material has not been transported.

rhyolite the aphanitic equivalent of granite.

rip rap broken rock used for such structures as breakwaters.

rock a natural solid composed of mineral grains, glass, or a combination of mineral grains and glass.

roundness a measurement of the relative sharpness of the edges and corners of a stone (see Figure 1.4).

rubble a detrital sediment consisting predominantly of angular fragments of cobble or boulder size.

sand a detrital sediment consisting predominantly of fragments between 1⁄16 and 2 mm in intermediate diameter.

sandstone a detrital sedimentary rock that is lithified sand.

schist a well-foliated metamorphic rock consisting of a significant percentage of one or more platy minerals such as micas and chlorites.

sediment loose rock material, consisting of fragments of preexisting rocks or chemically or biochemically precipitated minerals, that has accumulated on the earth's surface.

sedimentary used to describe rock consisting of consolidated sediment; also applied to processes related to the formation of such deposits.

serpentinite a metamorphic rock consisting largely of the mineral serpentine.

shale a sedimentary rock made up largely of clay particles arranged so that the rock has fissility (that is, will readily split parallel to bedding laminae).

sill a tabular, intrusive igneous mass that is parallel to the bedding or foliation of the surrounding country rocks (see Figure 3.3).

siltstone a detrital sedimentary rock consisting of silt size (1/256–1/16 mm) fragments.

sink hole a depression, typically nearly circular in plan, formed as the result of the collapse of a cavern roof.

slate a microcrystalline metamorphic rock that has rock cleavage—that is, it splits readily into slabs.

slickensides (also termed *slickensided surface*) smooth, commonly striated surface formed in response to slippage and friction between rocks within a fault zone.

soapstone a metamorphic rock made up largely or wholly of talc that does not have a preferred orientation.

solution chemical dissolving of a substance by a fluid.

species a biological subdivision based on the capability of its members to breed only with others of their kind and thereby produce offspring having the same capability.

specific gravity (SG) the ratio of the weight of a substance to the weight of an equal volume of water. It may be determined by weighing the object in air (W_a), weighing it immersed in water (W_w), and substituting the values in this equation:

$$SG = \frac{W_a}{W_a - W_w}.$$

specific mineral in an igneous rock, the mineral or minerals required by the definition of the rock.

speleothem overall name for cave deposits of chemical origin.

sphericity the degree to which a stone approaches the shape of a sphere; that is, the relationships of a stone's length, width, and thickness to each other (see Figure 1.5).

stalactite a roughly conical or icicle-shaped speleothem that hangs down from the roof of a cave.

stalagmite a roughly conical speleothem that protrudes upward from the floor of a cave.

stone a loose entity (larger than a sand grain) that is

made up of one or more minerals or rock materials and that has its loose aspect as a consequence of natural processes.

stratification the layering, or bedding, of sedimentary rock.

stratified drift glacially deposited material consisting of sorted and layered rock material.

streambed the channel in rock material through which a stream flows or has flowed.

striae (singular *stria*) grooves and scratches as, for example, on glacially faceted stones (see Figure 2.4*A*). Striae are sometimes called *striations*.

stylolite a thin, irregular, interdigitated contact seam—most commonly in limestone or dolostone—that is expressed on flat surfaces by a dark zigzag line (see Figure 3.15).

submetallic applied to minerals with lusters intermediate between metallic and nonmetallic.

syenite an igneous phanerite consisting of 65 to 90 percent alkali feldspar, 10 to 35 percent plagioclase, less than 5 percent quartz, and having a dark-colored varietal mineral content (typically hornblende) of between 10 and 35 percent.

tabular having a shape with one dimension much shorter than the other two.

tactite a metamorphic rock formed as the result of contact metamorphism involving noteworthy metasomatism.

tektite a glassy stone, typically globular or dumbbell shaped, probably formed by splashing of molten rock material formed as a result of meteoric impact (see Figure 2.22).

texture geometric interrelationships among constituent mineral grains in a rock.

thermal metamorphism metamorphism in response to elevated temperature.

thunderegg a term frequently applied to geodelike masses—typically with agate cores—that have formed within rhyolitic welded tuffs.

till unconsolidated material consisting of unsorted glacially deposited debris.

tillite consolidated glacial till.

trachyte the aphanitic equivalent of syenite.

trap rock a term frequently applied to basalt and diabase.

travertine calcite or aragonite deposited from solutions in caves (e.g., as speleothems) or around seeps or springs.

tuff overall term for lithified pyroclastic materials.

ultrabasic term for an igneous rock composed 90 percent or more of one or more of the iron-magnesium silicate minerals.

unakite epidote-rich granite (the rock is probably an epidotized granite).

vacuole a synonym for vesicle.

varietal mineral in an igneous rock, a mineral that is commonly, but not necessarily, present. The names of varietal minerals are generally used as adjectives in complete rock designations.

vein a mineral-filled fracture in rock (see Figure 3.19).

ventifact a stone that has been shaped, worn, or polished by windblown materials (see Figure 2.4*B*).

vesicle (adj. *vesicular*) a cavity, commonly spheroidal, formed by the expansion of a gas bubble during solidification of magma; cf. *amygdule*.

vitric (as applied to tuffs) consisting of fragments of volcanic glass, cf. *crystal* and *lithic*.

volcanic bomb a pyroclastic fragment that has been solidified from a fluid or partially fluid material while it was in flight through the air. Bombs typically have twisted shapes (see Figure 2.23).

vug a small cavity in rock. Vugs are typically lined with crystals.

weathering pertaining to the processes or products of chemical and physical changes that take place as the result of exposure to the elements.

xenolith an inclusion of country rock in an igneous mass (see Figure 3.3).

Suggested Readings

The references selected are listed under headings that correspond to the chapter titles and under an additional heading, General References. Annotations are given for some of the references; the titles of the others suffice to indicate their content. The reader also may wish to consult the *Encyclopedia Americana*, the *Encyclopaedia Britannica*, other encyclopedias, and *The Geology Merit Badge* used by the Boy Scouts of America. Periodicals of possible interest are listed at the end.

INTRODUCTION

Adams, F. D., 1938, *The Birth and Development of the Geological Sciences*. Williams and Wilkins, Baltimore (reprinted, 1954, by Dover, New York).

Berry, L. G., and Mason, B., 1983, *Mineralogy: Concepts, Descriptions, Determinations*, 2nd ed., revised by R. V. Dietrich. W. H. Freeman and Company, San Francisco.

Dietrich, R. V., 1969, *Mineral Tables, Hand-Specimen Properties of 1500 Minerals*. McGraw-Hill, New York.

Pough, F. H., 1976, *Field Guide to Rocks and Minerals*, 4th ed. Houghton Mifflin, Boston.

Zim, H., and Shaffer, P., 1957, *Rocks and Minerals* (Golden Nature Guide). Golden Press, New York.

OCCURRENCES

Dodd, R. T., 1986, *Thunderstones and Shooting Stars—The Meaning of Meteorites*. Harvard University Press, Cambridge, Massachusetts.

Dunbar, C. O., and Rodgers, J., 1957, *Principles of Stratigraphy*. Wiley, New York. Environments and general principles of sedimentary deposition are described.

Emery, K. O., 1965, "Organic transportation of marine sediments." In Hill, M. N. (ed.), *The Sea*, v. 1. Wiley Interscience, New York, p. 776–793. Examples of transportation of stones by plants and animals and criteria for recognizing rafting agents are described.

Flint, R. F., 1971, *Glacial and Quaternary Geology*. Wiley, New York. This book contains the most complete treatment of the Ice Age and materials deposited during that time.

Hunt, C. B., 1973, *Natural Regions of the United States and Canada*. W. H. Freeman and Company, San Francisco.

King, E. A., 1977, "The origin of tektites: A brief review." *American Scientist*, v. 65, p. 212–218.

Pettijohn, F. J., 1975, *Sedimentary Rocks*, 3rd ed. Harper & Row, New York. Chapter 6, p. 154–195, summarizes pre-1973 research on gravels and gravel deposits.

Sharp, R. P., and Carey, D. L., 1976, "Sliding stones, Racetrack Playa, California" *Geol. Soc. Amer. Bull.*, v. 87, p. 1704–1717.

Whitney, M. I., and Dietrich, R. V., 1973, "Ventifact sculpture by windblown dust." *Geol. Soc. Amer. Bull.*, v. 84, p. 2561–2582.

ROCK TYPES

Casanova, R. and Ratkevich, R. P., 1981, *An Illustrated Guide to Fossil Collecting*, 3rd revised ed. Naturegraph, Happy Camp, California.

Collinson, C. W., 1959, *Guide for Beginning Fossil Hunters*, revised ed. Illinois State Geol. Survey, Educ. Ser. 4.

Dietrich, R. V., 1974, *"Migmatites—A Resume." Jour. Geol. Educ.*, v. 22, p. 144–156. This is a summary about these "mixed rocks," which are not treated in most elementary text books.

Dietrich, R. V., and Skinner, B. J., 1979, *Rocks and Rock Minerals*. Wiley, New York.

Dietrich, R. V. and Wicander, R., 1983, *Minerals, Rocks, and Fossils A Self-Teaching Guide*. Wiley, New York.

McFall, R. P., and Wallin J., 1983, *Fossils for Amateurs*, 2nd ed. Van Nostrand Reinhold, New York.

Rhodes, F. H. T., Zim, H. S., and Shaffer, P. R., 1962, *Fossils* (Golden Nature Guide). Golden Press, New York.

Shimer, H. W., and Schrock, R. R., 1944, *Index Fossils of North America*. Wiley, New York. Although more than forty years old and containing some superseded designations, this book remains the paleontologist's standard for North America.

USES

Bateman, A. M., 1950, *Economic Mineral Deposits*, 2nd ed. Wiley, New York.

Bates, R., 1960, *Geology of the Industrial Rocks and Minerals*. Harper & Row, New York (reprinted, 1969, by Dover, New York). Good descriptions of nonmetallic deposits are given.

Dietrich, R. V., 1989, Rocks depicted in painting and sculpture. *Rocks & Minerals*, in press.

Dietrich, R. V., 1989, Rocks in haiku. *New Cicada Haiku*, in press.

Hackensmith, C. W., 1966, *History of Physical Education*. Harper & Row, New York.

Hay, J., 1985, *Kernels of Energy, Bones of Earth: The Rock in Chinese Art*. China Institute in America, New York.

Lippard, L. R., 1983, *Overlay: Contemporary Art and the Art of Prehistory*. Pantheon, New York.

Schmidt, C. F., 1966, *Cobblestone Masonry*. Private printing, Scottsville, New York.

VonFrisch, K. and vonFrisch, O., 1974, *Animal Architecture*. Harcourt Brace Jovanovich, New York (reprinted, 1983, by Van Nostrand Reinhold, New York).

West, J. M., 1971, *How to Mine and Prospect for Placer Gold*. U.S. Bureau of Mines, Information Circular 8517. Methods for panning, among other things, are clearly outlined.

Willey, R. R., 1987, *The Tucson Meteorites*. Smithsonian Institution Press, Washington, D. C.

COLLECTING STONES

The American Association of Petroleum Geologists (P.O. Box 979, Tulsa, Oklahoma 74101) publishes geological highway maps for different regions of the United States.

Dietrich, R. V., 1989, Imagine: Another mimetolith. *Rocks & Minerals*, v. 64, p. 149–152.

Dietrich, R. V. and Wicander, R., 1983, *Minerals, Rocks, and Fossils: A Self-Teaching Guide*. Wiley, New York.

Hyde, C. S., and Matthews, R. A., 1977, *The Complete Book of Rock Tumbling*. Chilton Book Co., Radnor, Pennsylvania.

GENERAL REFERENCES

American Geological Institute, 1984, *Dictionary of Geological Terms* revised ed. Anchor Press/Doubleday, Garden City, New York.

Bates, R. L. and Jackson, J. A. (eds.), 1987, *Glossary of Geology*, 3rd ed. American Geological Institute, Alexandria, Virginia.

Gilluly, J., Waters, A. C., and Woodford, A. O., 1975, *Principles of Geology*, 4th ed. W. H. Freeman and Company, San Francisco, 527 p.

Noe-Nygaard, A., 1960, *Strandsten*. Gyldendal, Copenhagen, Denmark, 103 p. This book, in Danish, is about stones found on the beaches of Denmark. The accompanying pen and ink sketches of stones by Gunni Jørgensen are outstanding.

Press, F., and Siever, R., 1986, *Earth*, 4th ed. W. H. Freeman and Company, New York.

Rhodes, F. H., 1971, *Geology* (Golden Nature Guide). Golden Press, New York.

Skinner, B. J., and Porter, S. C., 1987, *Physical Geology*. Wiley, New York. One of several introductory geology textbooks, this book contains much background information. The books by Gilluly et al. and by Press and Siever, which are listed above, are two other standard texts with somewhat different approaches.

MAGAZINES

The following periodicals, although dealing more specifically with minerals, rocks, or fossils (as opposed to stones per se), may be of interest. Those listed have been selected because of the type (including illustration content) or level of coverage.

Lapidary Journal: Box 80937, San Diego, California, 92138.

Rock & Gem: Miller Inc., 2660 E. Main St., Ventura, California 93003.

Rocks & Minerals: Room 500, 4000 Albemarle Street, N. W., Washington, DC. 20016.

The Mineralogical Record: 4631 Paseo Tubutama, Tucson, Arizona, 85740.

Lapis: Christian Weise Verlag, Oberanger 6, D-8000 Munchen 2, West Germany.

Mineralienfreund: Postfach 219, 6460 Altdorf-Uri, Switzerland.

Mineralien Magazin: KOSMOS-Verlag, Abt. 16, Postfach 640, D-7000 Stuttgart, West Germany.

Mineraux et Fossiles: 12 rue Poincare, 558700 Revigny, France.

Monde et Mineraux: 17 rue Guersant, 75017, Paris, France.

Index

Page references to illustrations are followed by asterisks (*). Page references to words in the glossary are given in **boldface type**.